Lecture Notes in Computer Science

Lecture Notes in Artificial Intelligence   **13934**

Founding Editor

Jörg Siekmann

Series Editors

Randy Goebel, *University of Alberta, Edmonton, Canada*
Wolfgang Wahlster, *DFKI, Berlin, Germany*
Zhi-Hua Zhou, *Nanjing University, Nanjing, China*

The series Lecture Notes in Artificial Intelligence (LNAI) was established in 1988 as a topical subseries of LNCS devoted to artificial intelligence.

The series publishes state-of-the-art research results at a high level. As with the LNCS mother series, the mission of the series is to serve the international R & D community by providing an invaluable service, mainly focused on the publication of conference and workshop proceedings and postproceedings.

Dominik Dürrschnabel ·
Domingo López Rodríguez

Editors

Formal Concept Analysis

17th International Conference, ICFCA 2023
Kassel, Germany, July 17–21, 2023
Proceedings

 Springer

Editors
Dominik Dürrschnabel [ID]
Universität Kassel
Kassel, Germany

Domingo López Rodríguez [ID]
Universidad de Málaga
Málaga, Spain

ISSN 0302-9743   ISSN 1611-3349 (electronic)
Lecture Notes in Artificial Intelligence
ISBN 978-3-031-35948-4   ISBN 978-3-031-35949-1 (eBook)
https://doi.org/10.1007/978-3-031-35949-1

LNCS Sublibrary: SL7 – Artificial Intelligence

This Springer imprint is published by the registered company Springer Nature Switzerland AG
The registered company address is: Gewerbestrasse 11, 6330 Cham, Switzerland

Preface

The field of Formal Concept Analysis (FCA) originated in the 1980s in Darmstadt, Germany, as a subfield of mathematical order theory, building on earlier research and developments by other groups. The original motivation for FCA was to explore complete lattices as lattices of concepts, drawing inspiration from both Philosophy and Mathematics. Over time, FCA has evolved into a broad research area with applications far beyond its initial focus, including but not limited to logic, knowledge representation, unsupervised learning, data mining, human learning, and psychology.

This volume features the contributions accepted for the 17th International Conference on Formal Concept Analysis (ICFCA 2023), held during July 17–21, 2023, at the University of Kassel, Germany. The International Conference on Formal Concept Analysis serves as a platform for researchers from FCA and related disciplines to showcase and exchange their research findings. Since its inaugural edition in 2003 in Darmstadt, Germany, ICFCA has been held annually in multiple locations across Europe, Africa, America, and Australia. In 2015, ICFCA became a biennial event to alternate with the Conference on Concept Lattices and Their Applications (CLA).

The previous edition, ICFCA 2021, organized by the Université de Strasbourg in France, was held online due to the COVID-19 pandemic. Now that conditions are beginning to return to the pre-pandemic situation, this edition will once again offer face-to-face opportunities as well as an online mode for the audience, making this the first hybrid ICFCA edition.

This year's ICFCA received 19 submissions from authors from ten countries. At least two different program committee members and one member of the editorial board reviewed each submission. Reviewing was done in a single-blinded fashion. Thirteen high-quality papers were selected for publication in this volume. This represents an acceptance rate of 68%.

The research part of this volume is divided into two distinct sections. First, in "Theory", we have collected papers that propose and review advances in theoretical aspects of FCA. The second section, "Applications and Visualization", consists of advances that deal with new algorithms, applications, and different approaches to visualization techniques.

In addition to the regular contributions, this volume also includes the abstracts of the six invited talks by outstanding researchers whom we were pleased to welcome to ICFCA 2023. In detail, we were pleased to host the following talks:

- *How to Visualize Sets and Set Relations* by Oliver Deussen, Visual Computing, University of Konstanz, Germany
- *Tangles: from Wittgenstein to graph minors and back* by Reinhard Diestel, Discrete Mathematics, University of Hamburg, Germany
- *Formal Concept Analysis in Boolean Matrix Factorization: Algorithms and Extensions to Ordinal and Fuzzy-Valued Data* by Jan Konečný, Department of Computer Science, Palacký University Olomouc, Czech Republic

- *On the φ-degree of inclusion* by Manuel Ojeda Aciego, Departmento de Matemática Aplicada, Universidad de Málaga, Spain
- *Logical foundations of categorization theory* by Alessandra Palmigiano, Department of Ethics, Governance and Society, Vrije Universiteit Amsterdam, The Netherlands
- *Modern Concepts of Dimension for Partially Ordered Sets* by William T. Trotter, School of Mathematics, Georgia Tech Institute, USA

Moreover, due to the recent computation of the ninth Dedekind number using techniques from FCA, this edition also included a late-breaking result talk:

- *Breaking the Barrier: A Computation of the Ninth Dedekind Number* by Christian Jäkel, Methods of Applied Algebra, Dresden University of Technology, Germany

We are deeply grateful to all the authors who submitted their papers to ICFCA 2023 as a venue to present their work. Our sincere appreciation goes to the members of the Editorial Board and Program Committee, as well as all the additional reviewers, whose timely and thorough reviews enabled the fruitful discussions of the high-quality papers during the conference. We would also like to express our gratitude to the local organizers, who were always quick to solve any problems that arose. We are very grateful to Springer for supporting the International Conference on Formal Concept Analysis and to the Department of Electrical Engineering and Computer Science and to the Research Center for Information System Design of the University of Kassel for hosting the event.

July 2023

Dominik Dürrschnabel
Domingo López-Rodríguez
Gerd Stumme

Organization

Executive Committee

Conference Chair

Gerd Stumme   Universität Kassel, Germany

Local Organization Committee

Tobias Hille   Universität Kassel, Germany
Johannes Hirth   Universität Kassel, Germany
Maximilian Stubbemann   Universität Kassel, Germany

Program and Conference Proceedings

Program Chairs

Dominik Dürrschnabel   Universität Kassel, Germany
Domingo López-Rodríguez   Universidad de Málaga, Spain

Editorial Board

Jaume Baixeries   Polytechnic University of Catalonia, Spain
Peggy Cellier   INSA Rennes, France
Sebastien Ferrè   Université de Rennes 1, France
Bernhard Ganter   Technische Universität Dresden, Germany
Tom Hanika   Universität Kassel, Germany
Dmitry Ignatov   Higher School of Economics, Moscow, Russia
Mehdi Kaytoue   INSA-Lyon, LIRIS, France
Sergei Kuznetsov   Higher School of Economics, Moscow, Russia
Leonard Kwuida   Bern University of Applied Sciences, Switzerland
Florence Le Ber   Université de Strasbourg, France
Rokia Missaoui   Université du Québec en Outaouais, Canada
Amedeo Napoli   LORIA, Nancy, France
Sergei Obiedkov   Higher School of Economics, Moscow, Russia
Manuel Ojeda-Aciego   Universidad de Málaga, Spain

Uta Priss	Ostfalia University of Applied Sciences, Germany
Christian Sacarea	Babes-Bolyai University of Cluj-Napoca, Romania
Stefan E. Schmidt	Technische Universität Dresden, Germany
Barş Sertkaya	Frankfurt University of Applied Sciences, Germany
Petko Valtchev	Université du Québec à Montréal, Canada

Program Committee

François Brucker	École Centrale de Marseille, France
Victor Codocedo	Universidad Técnica Federico Santa María, Chile
Pablo Cordero	Universidad de Málaga, Spain
Diana Cristea	Babes-Bolyai University Cluj-Napoca, Romania
Christophe Demko	Université de La Rochelle, France
Jean Diatta	Université de la Réunion, France
Stephan Doerfel	Micromata GmbH, Germany
Xavier Dolques	Université de Strasbourg, France
Marianne Huchard	Université de Montpellier, France
Jan Konecny	Palacky University of Olomouc, Czech Republic
Wilfried Lex	Universität Clausthal, Germany
Jesús Medina	Cadiz University, Spain
Engelbert Mephu Nguifo	Université de Clermont Ferrand 2, France
Jan Outrata	Palacky University of Olomouc, Czech Republic
Henry Soldano	Université Paris 13, France
Laszlo Szathmary	University of Debrecen, Hungary
Andreja Tepavcevic	University of Novi Sad, Serbia
Martin Trnecka	Palacky University of Olomouc, Czech Republic

Additional Reviewers

Mohamed Hamza Ibrahim	Université du Québec en Outaouais, Canada

Abstracts of Invited Talks

How to Visualize Sets and Set Relations

Oliver Deussen

Visual Computing, University of Konstanz, Germany

In his talk he will focus on one particular area of information visualization: the visualization of sets and set relations (known by Venn and Euler diagrams). Unfortunately, such diagrams do not scale, they do not work for larger set numbers. Furthermore, the human ability to understand such data is limited, therefore his group developed perceptually-driven methods to improve these visualizations. Based on a review of existing methods in information visualization he will present new approaches.

Tangles: from Wittgenstein to Graph Minors and Back

Reinhard Diestel

Discrete Mathematics, University of Hamburg, Germany

Tangles, a central notion from the Robertson-Seymour theory of graph minors, have recently been axiomatised in a way that allows us to capture fuzzy phenomena of connectivity and cohesion, such as mindsets in sociology, topics in text analysis, or Wittgenstein's family resemblances in linguistics, in an entirely formal and precise way.

This talk will offer a non-technical introduction to tangle theory that highlights its various potential applications. Participants interested in trying them out in their own field of interest can obtain well-documented code from us to facilitate this. I look forward to stimulating co-operation with the FCA community!

Formal Concept Analysis in Boolean Matrix Factorization: Algorithms and Extensions to Ordinal and Fuzzy-Valued Data

Jan Konečný

Department of Computer Science, Palacký University Olomouc, Czech Republic

It has been nearly fifteen years since Belohlavek and Vychodil first highlighted the usefulness of Formal Concept Analysis (FCA) in Boolean Matrix Factorization (BMF) and introduced the initial FCA-based algorithms for BMF. This work sparked a thriving research direction within our department that persists to this day. The talk provides an overview of our ongoing research efforts, with a particular emphasis on the application of FCA in the development of algorithms for BMF. Moreover, the progress made in extending these techniques to the factorization of matrices containing ordinal and fuzzy-valued data will be discussed.

On the φ-Degree of Inclusion

Manuel Ojeda Aciego

Departamento de Matemática Aplicada, Universidad de Málaga, Spain

The notion of inclusion is a cornerstone in set theory and therefore its generalisation in fuzzy set theory is of great interest. The functional degree (or φ-degree) of inclusion is defined to represent the degree of inclusion between two L-fuzzy sets in terms of a mapping that determines the minimal modifications required in one L-fuzzy set to be included in another in the sense of Zadeh. Thus, this notion differs from others existing in the literature because the φ-degree of inclusion is considered as a mapping instead of a value in the unit interval. We show that the φ-degree of inclusion satisfies versions of many common axioms usually required for inclusion measures in the literature.

Considering the relationship between fuzzy entropy and Young's axioms for measures of inclusion, we also present a measure of entropy based on the φ-degree of inclusion that is consistent with the axioms of De Luca and Termini. We then continue to study the properties of the φ-degree of inclusion and show that, given a fixed pair of fuzzy sets, their φ-degree of inclusion can be linked to a fuzzy conjunction that is part of an adjoint pair. We also show that when this pair is used as the underlying structure to provide a fuzzy interpretation of the modus ponens inference rule, it provides the maximum possible truth value in the conclusion among all those values obtained by fuzzy modus ponens using any other possible adjoint pair. Finally, we will focus on current work on the integration of the φ-degree of inclusion with FCA.

Logical Foundations of Categorization Theory

Alessandra Palmigiano

Department of Ethics, Governance and Society, Vrije Universiteit Amsterdam,
The Netherlands

Categories are cognitive tools that humans use to organize their experience, understand and function in the world, and understand and interact with each other, by grouping things together which can be meaningfully compared and evaluated. They are key to the use of language, the construction of knowledge and identity, and the formation of agents' evaluations and decisions. Categorization is the basic operation humans perform e.g. when they relate experiences/actions/objects in the present to ones in the past, thereby recognizing them as instances of the same type. This is what we do when we try to understand what an object is or does, or what a situation means, and when we make judgments or decisions based on experience. The literature on categorization is expanding rapidly in fields ranging from cognitive linguistics to social and management science to AI, and the emerging insights common to these disciplines concern the dynamic essence of categories, and the tight interconnection between the dynamics of categories and processes of social interaction. However, these key aspects are precisely those that both the extant foundational views on categorization struggle the most to address. In this lecture, we will discuss a logical approach, semantically based on formal contexts and their associated concept lattices, which aims at creating an environment in which these three cognitive processes can be analyzed in their relationships to one another, and propose several research directions, developing which, novel foundations of categorization theory can be built.

Modern Concepts of Dimension for Partially Ordered Sets

William T. Trotter

School of Mathematics, Georgia Tech Institute, USA

Partially ordered sets (posets) are useful models for rankings on large data sets where linear orders may fail to exist or may be extremely difficult to determine. Many different parameters have been proposed to measure the complexity of a poset, but by far the most widely studied has been the Dushnik-Miller notion of dimension. Several alternate forms of dimension have been proposed and studied, and the pace of research on this theme has accelerated in the past 10 years. Among these variants are Boolean dimension, local dimension, fractional dimension and fractional local dimension. The last in this list holds promise for applications since it provides an answer to the natural question: How can we determine a partial order given only piecemeal observations made, often independently, and typically with conflicts, on extremely large populations? As an added bonus, this line of research has already produced results that have intrinsic mathematical appeal.

In this talk, I will introduce the several concepts of dimension, outline some key results and applications for each, and close with more detailed comments on fractional local dimension.

Latebreaking Result Talk

Breaking the Barrier: A Computation of the Ninth Dedekind Number

Christian Jäkel

Technische Universität Dresden, Germany

The Dedekind numbers are a fast-growing sequence of integers which are exceptionally hard to compute. Introduced in 1897 by Richard Dedekind, the nth Dedekind number represents the size of the free distributive lattice with n generators, the number of antichains of subsets of an n-element set, and the number of abstract simplicial complexes with n elements. It is also equal to the number of Boolean functions with n variables, composed only of "and" and "or" operators.

After three decades of being an open problem, I will present my successful approach to compute the ninth Dedekind number in this talk. Inspired by Formal Concept Analysis, I developed complexity reduction techniques to tackle this problem. In addition, I discovered formulas that facilitate the efficient computation on GPUs, resulting in significantly faster runtimes compared to conventional CPU computing.

Contents

Theory

Approximating Fuzzy Relation Equations Through Concept Lattices

David Lobo[ID], Víctor López-Marchante[ID], and Jesús Medina[(✉)][ID]

Department of Mathematics, University of Cádiz, Cádiz, Spain
{david.lobo,victor.lopez,jesus.medina}@uca.es

Abstract. Fuzzy relation equations (FRE) is a formal theory broadly studied in the literature and applied to decision making, optimization problems, image processing, etc. It is usual that the initial data contains uncertain, imperfect or incomplete information, which can imply, for instance, the existence of inconsistencies. As a consequence, the FRE that arises from the data may be unsolvable. Taking advantage of the relationship between FRE and concept lattices, this paper is focused on three mechanisms for approximating unsolvable FRE. Several properties have been introduced and different distances for determining the best approximation are considered and applied to an example.

Keywords: Fuzzy Sets · Fuzzy Relation Equation · Property-oriented concept lattice · Adjoint triples

1 Introduction

Formal concept analysis (FCA) has interacted with other formal tools and approaches in order to obtain robust and trustworthy intelligent systems, such as rough set theory [10,22], machine learning [18,21,28], graph theory [3,17,26], etc. From these interactions, this paper will be focused on the relationship with fuzzy relation equations (FRE), which were introduced by Elie Sanchez in 1976 [25]. This theory has been broadly studied in the literature by many authors [11,12,24,27], spreading to fields like decision making [4], bipolarity [5,6], optimization [2] and image processing [1].

The relationship between FRE and the isotone variants of FCA, that is, the property-oriented concept lattices and object-oriented concept lattices [22], was originally presented by Díaz-Moreno and Medina in [14]. This relation was

Partially supported by the 2014–2020 ERDF Operational Programme in collaboration with the State Research Agency (AEI) in project PID2019-108991GB-I00, with the Ecological and Digital Transition Projects 2021 of the Ministry of Science and Innovation in project TED2021-129748B-I00, and with the Department of Economy, Knowledge, Business and University of the Regional Government of Andalusia in project FEDER-UCA18-108612, and by the European Cooperation in Science & Technology (COST) Action CA17124.

D. Dürrschnabel and D. López Rodríguez (Eds.): ICFCA 2023, LNAI 13934, pp. 3–16, 2023.
https://doi.org/10.1007/978-3-031-35949-1_1

intensified with the generalization of FRE to the multi-adjoint paradigm in [13, 15]. A complete study on the computation of the whole solution set of a solvable FRE is shown in [15], where it is characterized in terms of an associated multi-adjoint property-oriented concept lattice.

Multi-adjoint relation equations (MARE) [13] provide a large flexibility for setting the equations, due to the generality of the underlying algebraic structure and the possibility of considering different conjunctors for each variable depending on its nature.

A given FRE might be not solvable because of different reasons. For example, because of an excess of imposed restrictions, the consideration of constrains that cannot be satisfied or the imposition of two incompatible conditions. It is therefore necessary to have mechanisms that allow to transform a given unsolvable FRE into a solvable one, through slight changes. We will refer to this process as approximating an unsolvable FRE to a solvable one. In addition, it is important that the FRE approximation procedures can be easily interpretable, in the sense that the effect that they have on the original equation when applied are known and controlled.

This paper will analyze three mechanisms with different philosophies for approximating an unsolvable FRE and their main features will be discussed. Moreover, several distances for measuring the change made by the approximation processes and the comparison among them on an example will also be presented. The three mechanisms arise thanks to the relationship between FRE and property-oriented concept lattices and object-oriented concept lattices [13]. The first one was recently introduced in [20] and it is based on attribute reduction in FCA [8]. The attribute reduction results in FCA were translated to property-oriented and object-oriented concept lattices and applied to FRE. As a first consequence, a mechanism for simplifying FRE without removing relevant information was introduced, and this reduction also offers the possibility of approximating FRE. The other two methods were introduced in [4], where a comparison between FRE and logic programming was done. In this paper, we present these procedures in a general framework and several properties will be studied.

The paper is structured as follows. Section 2 includes the algebraic structure considered in the paper, and several basic notions and results on FRE and concept lattices. Section 3 presents the three mechanisms to approximate FRE and analyzes diverse properties. Then, Sect. 4 considers four distances in order to compare the obtained approximated FRE on an example. Finally, some conclusions and prospects for future work are given.

2   Preliminaries

The algebraic structures considered in the multi-adjoint frameworks are based on adjoint triples. These operators generalize left-continuous t-norms and its residuated implications, where the conjunctors do not need to be commutative or associative. Consequently, two different residuated implications arise associated with these conjunctions.

Definition 1 ([9]). *Let* $(P_1, \preceq_1)$, $(P_2, \preceq_2)$, $(P_3, \preceq_3)$ *be posets and* $\&\colon P_1 \times P_2 \to P_3$, $\nearrow\colon P_3 \times P_2 \to P_1$, $\nwarrow\colon P_3 \times P_1 \to P_2$ *mappings, then* $(\&, \nearrow, \nwarrow)$ *is an* adjoint triple *with respect to* P_1, P_2, P_3 *if*

$$x \preceq_1 z \nearrow y \quad \text{iff} \quad x \& y \preceq_3 z \quad \text{iff} \quad y \preceq_2 z \nwarrow x$$

for each $x \in P_1$, $y \in P_2$, $z \in P_3$.

The Gödel, Łukasiewicz and product t-norms, together with their residuated implications are natural examples of adjoint triples. Moreover, other non-associative and non-commutative operators, such as $\&\colon [0, 1] \times [0, 1] \to [0, 1]$, defined as $x \& y = x^2 y$, for all $x, y \in [0, 1]$, can be considered [16]. Another kind of interesting operators are the discretization of the t-norms [23], which are the real operators used in the computations due to the fact that every numerical computation software uses a finite number of decimals. Recall that the Gödel t-norm $\&_G$ is defined as $x \&_G y = \min\{x, y\}$ for all $x, y \in [0, 1]$; the product t-norm $\&_p$ is defined as $x \&_P y = xy$ for all $x, y \in [0, 1]$ and the Łukasiewicz t-norm $\&_L$ is defined as $x \&_L y = \max\{0, x + y - 1\}$ for all $x, y \in [0, 1]$.

We will consider the regular partitions of $[0, 1]$ into m pieces, which will be denoted as $[0, 1]_m$. For example, $[0, 1]_4 = \{0, 0.25, 0.5, 0.75, 1\}$ divides the unit interval into four pieces. A discretization of a t-norm $\&\colon [0, 1] \times [0, 1] \to [0, 1]$ is the operator $\overline{\&}\colon [0, 1]_m \times [0, 1]_n \to [0, 1]_k$, where $n, m, k \in \mathbb{N}$, defined as:

$$x \,\overline{\&}\, y = \frac{\lceil k \cdot (x \& y) \rceil}{k}$$

for all $x \in [0, 1]_m$ and $y \in [0, 1]_n$, where $\lceil \_ \rceil$ is the ceiling function. The discretization of the corresponding residuated implications $\nearrow\colon [0, 1]_k \times [0, 1]_n \to [0, 1]_m$ and $\nwarrow\colon [0, 1]_k \times [0, 1]_m \to [0, 1]_n$ is defined as:

$$z \,\overline{\nearrow}\, y = \frac{\lfloor m \cdot (z \leftarrow y) \rfloor}{m} \qquad z \,\overline{\nwarrow}\, x = \frac{\lfloor n \cdot (z \leftarrow x) \rfloor}{n}$$

for all $z \in [0, 1]_k$ and $y \in [0, 1]_n$, where $\lfloor \_ \rfloor$ is the floor function and $\leftarrow$ is the residuated implication of the t-norm $\&$. We have that $(\overline{\&}, \overline{\nearrow}, \overline{\nwarrow})$ is an adjoint triple with respect to $[0, 1]_n$, $[0, 1]_m$ and $[0, 1]_k$ [7, 23].

The concept of multi-adjoint property-oriented frame was introduced in [22] with the aim of allowing the use of several adjoint triples. For computational reasons, we will assume that two of the posets are lattices.

Definition 2 ([22]). *Let* $(L_1, \preceq_1)$, $(L_2, \preceq_2)$ *be two lattices,* $(P, \preceq_3)$ *a poset and* $\{(\&_i, \nearrow^i, \nwarrow_i) \mid i \in \{1, \ldots, n\}\}$ *a set of adjoint triples with respect to* P, L_2, L_1. *The tuple*

$$(L_1, L_2, P, \preceq_1, \preceq_2, \preceq_3, \&_1, \nearrow^1, \nwarrow_1, \ldots, \&_n, \nearrow^n, \nwarrow_n)$$

is called multi-adjoint property-oriented frame.

The following definition presents the notion of context.

Definition 3 ([22]). *Let* $(L_1, L_2, P, \&_1, \ldots, \&_n)$ *be a multi-adjoint property-oriented frame. A* multi-adjoint context *is a tuple* (A, B, R, σ) *where A and B are non-empty sets, $R\colon A \times B \to P$ is a fuzzy relation and $\sigma\colon A \times B \to \{1, \ldots, n\}$ is a mapping.*

Given a multi-adjoint frame and a context, two mappings can be defined between the fuzzy subsets of *attributes* A, that is $L_1^A = \{f \mid f\colon A \to L_1\}$, and the fuzzy subsets of *objects* B, that is $L_2^B = \{g \mid g\colon B \to L_2\}$. Specifically, the mappings $\uparrow^\pi\colon L_2^B \to L_1^A$ and $\downarrow^N\colon L_1^A \to L_2^B$ are defined as

$$g^{\uparrow_\pi}(a) = \bigvee{}_1 \left\{ R(a, b) \,\&_{\sigma(a,b)}\, g(b) \mid b \in B \right\} \tag{1}$$

$$f^{\downarrow^N}(b) = \bigwedge{}_2 \left\{ f(a) \,\nwarrow_{\sigma(a,b)}\, R(a, b) \mid a \in A \right\} \tag{2}$$

for all $f \in L_1^A$ and $g \in L_2^B$, where $\bigvee_1$ and $\bigwedge_2$ represent the suprema and infima of $(L_1, \preceq_1)$ and $(L_2, \preceq_2)$, respectively. The pair of mappings $(\uparrow^\pi, \downarrow^N)$ forms an isotone Galois connection [22]. This leads to the definition of *multi-adjoint property-oriented concept lattice.*

Consider the order relation $\preceq_{\pi N}$ defined as $\langle g_1, f_1 \rangle \preceq_{\pi N} \langle g_2, f_2 \rangle$ if and only if $f_1 \preceq_1 f_2$, or equivalently, if and only if $g_1 \preceq_2 g_2$. The multi-adjoint property-oriented concept lattice associated with the multi-adjoint property-oriented frame is given by

$$\mathcal{M}_{\pi N}(A, B, R, \sigma) = \left\{ \langle g, f \rangle \in L_2^B \times L_1^A \mid g = f^{\downarrow^N},\, f = g^{\uparrow_\pi} \right\} \tag{3}$$

The set $\mathcal{M}_{\pi N}$ endowed with the order $\preceq_{\pi N}$ is a complete lattice [22]. Given a concept $\langle g, f \rangle$, the mapping g is called *extent* of the concept and f is called the *intent* of the concept. The set of intents is denoted as $\mathcal{I}(\mathcal{M}_{\pi N})$ and the extents are denoted as $\mathcal{E}(\mathcal{M}_{\pi N})$.

Next, several definitions and results concerning multi-adjoint relation equation (MARE) will be recalled. For more details see [13]. A MARE is an equation of the form $R \odot S = T$, where R, S and T are fuzzy relations, R or S is unknown, and $\odot$ is a sup-composition operator. There exist different composition operators $\odot$ between fuzzy relations, in this paper we will consider the following one.

Definition 4 ([13]). *Let* U, V, W *be sets,* $(P_1, \preceq_1), (P_2, \preceq_2), (P_3, \preceq_3)$ *posets,* $\{(\&_i, \nearrow^i, \nwarrow_i) \mid i \in \{1, \ldots, n\}\}$ *a set of adjoint triples with respect P_1, P_2, P_3,* $\sigma\colon V \to \{1, \ldots, n\}$ *a mapping, and $R \in P_1^{U \times V}$, $S \in P_2^{V \times W}$, $T \in P_3^{U \times W}$ three fuzzy relations. If P_3 is a complete lattice, the operator $\odot_\sigma\colon P_1^{U \times V} \times P_2^{V \times W} \to P_3^{U \times W}$ defined as*

$$R \odot_\sigma S(u, w) = \bigvee{}_3 \{ R(u, v) \&_{\sigma(v)} S(v, w) \mid v \in V \} \tag{4}$$

is called sup-$\&_\sigma$-composition.

This sup-composition operator is used in the definition of the MARE analyzed in this paper.

Definition 5 ([13]). *A MARE with sup-&$_\sigma$-composition is an equality of the form*

$$R \odot_\sigma X = T \tag{5}$$

where X is an unknown fuzzy relation.

A MARE is *solvable* if there exists at least one solution, that is, a relation X exists satisfying (5). Otherwise, we say that the MARE is *unsolvable*. A dual equation can be developed if the relation R is the unknown relation in the equation $R \odot_\sigma S = T$.

If P_2 is a lower-semilattice and the MARE (5) is solvable, then its greatest solution exists and can be computed as follows.

Proposition 1 ([13]). *Let $R \odot_\sigma X = T$ be a solvable MARE and $(P_2, \preceq_2)$ a lower-semilattice. Its greatest solution $\overline{X}$ is given by*

$$\overline{X}(v,w) = \bigwedge\nolimits_2 \{T(u,w) \nwarrow_{\sigma(v)} R(u,v) \mid u \in U\}$$

for each $(v,w) \in V \times W$

In [13,15], diverse results related to the solvability and the computation of the solution set of a MARE were introduced based on its relationship with the isotone variants of FCA. Next, the most relevant ones in the scope of this paper will be recalled. First of all, the following definition assigns a context to a MARE.

Definition 6 ([13]). *The* multi-adjoint context associated with the MARE $R\odot_\sigma X = T$ *is the property-oriented multi-adjoint context (U, V, R, σ).*

The following results are based on the notion of multi-adjoint concept lattice. The resolution procedure consists of associating a MARE with a context and, consequently, a multi-adjoint property-oriented concept lattice is associated with it too.

The concept lattice $(\mathcal{M}_{\pi N}(U, V, R, \sigma), \preceq_{\pi N})$ of the context associated with a MARE characterizes its solvability.

Proposition 2 ([13]). *Let (U, V, R, σ) be the multi-adjoint context associated with a MARE $R \odot_\sigma X = T$ and $(\mathcal{M}_{\pi N}, \preceq_{\pi N})$ the concept lattice associated with that context. Then $R \odot_\sigma X = T$ is solvable if and only if $T_w \in \mathcal{I}(\mathcal{M}_{\pi N})$ for all $w \in W$, where $T_w(u) = T(u, w)$, for all $u \in U$, $w \in W$.*

The solution set of a MARE can be characterized in terms of its associated context as was proved in [15]. Moreover, attribute reduction in FCA allows to reduce a MARE, thus leading to the notion of reduced MARE.

Definition 7 ([20]). *Let $Y \subseteq U$ and consider the relations $R_Y = R_{|Y \times V}$, $T_Y = T_{|Y \times W}$. The MARE $R_Y \odot_\sigma X = T_Y$ is called Y-reduced MARE of $R \odot_\sigma X = T$.*

Reducing a MARE in a consistent set/reduct preserves its solution set. The following result is one of the most relevant ones introduced in [20].

Theorem 1 ([20]). *Let $R \odot_\sigma X = T$ be a solvable MARE and Y a consistent set of its associated context (U, V, R, σ). The Y-reduced equation of $R \odot_\sigma X = T$ is solvable. In addition, $X \in L_2^{V \times W}$ is a solution of the Y-reduced equation if and only if it is a solution of the whole equation.*

3   Approximating Fuzzy Relation Equations

This section presents three different mechanisms for approximating unsolvable FRE, which have been adapted from [4,20]. Although the results are focused on the multi-adjoint framework, they can straightforwardly translated to other more particular settings, such as the residuated case and max-min FRE. The main common feature of the three approximation techniques is that they have been reached taking advantage of the relationship between FRE and concept lattices [13,15].

In the following, let us consider three sets U, V, W fixed and a property-oriented multi-adjoint framework

$$(L_1, L_2, P, \preceq_1, \preceq_2, \preceq, \&_1, \nearrow^1, \nwarrow_1, \dots, \&_n, \nearrow^n, \nwarrow_n)$$

A MARE is the equation

$$R \odot_\sigma X = T \tag{6}$$

where $R \in P^{U \times V}$, $T \in L_1^{U \times W}$ and $X \in L_2^{V \times W}$, with X being unknown.

The intuition behind the first approximation procedure is to eliminate several of the rows of an unsolvable MARE to recover its solvability. If this happens, it can be reasoned that the inconsistencies that gave rise to the unsolvability were due to the eliminated rows. However, as with reduction procedures, removing too many rows from the equation can lead to a loss of relevant information. In fact, eliminating all rows always leads to a trivially solvable equation, but no information from the original equation is preserved in that case. Therefore, we will use the approach of Theorem 1, which states that reducing a solvable equation considering a consistent set does not cause loss of information.

Based on the previous considerations, we will search for a consistent set of attributes Y of its associated context, such that the Y-reduced equation will be solvable, which means that all inconsistencies have been eliminated without losing relevant information. Moreover, in order to eliminate redundant information as well, we will always take into account reducts, which gives rise to the following definition.

Definition 8. *Let $R \odot_\sigma X = T$ be an unsolvable MARE and (U, V, R, σ) its associated context. We say that a reduct $Y \subseteq U$ of (U, V, R, σ) is feasible if the Y-reduced equation $R_Y \odot_\sigma X = T_Y$ is solvable.*

Note that every consistent set contains at least one reduct. Thus, in order to include as little redundant information as possible, feasible reducts will be considered from now on. Given a subset of attributes $Y \subseteq U$, we will use $(\uparrow_\pi^Y, \downarrow_Y^N)$ to denote the Galois connection associated with the context $(Y, V, R_Y, \sigma_{|Y \times V})$.

Suppose that an unsolvable MARE $R \odot_\sigma X = T$ admits a feasible reduct $Y \subseteq U$. Since the equation $R_Y \odot_\sigma X = T_Y$ is solvable, it is clear that the inconsistencies of the original MARE correspond to the rows that have been eliminated. Now, the question that naturally arises is whether it is possible to redefine the independent term of those rows and so, preserving the dimension of the original MARE but with a modified independent term offering a solvable equation without the need to eliminate any row. Thus, the proposed approximation method will modify the values of the independent term associated with attributes of the set $U \setminus Y$, so that the resulting MARE is solvable. The following result proves that, given a feasible reduct, it is always possible to approximate an unsolvable MARE in this way.

Theorem 2 ([20]). *Let* $R \odot_\sigma X = T$ *be an unsolvable MARE,* Y *a feasible reduct of its associated context* (U, V, R, σ), *and* T^{*Y} *the relation defined as* $T^{*Y}(u, w) = (T_Y)_w^{\downarrow_Y^N \uparrow_\pi}(u)$ *for all* $(u, w) \in U \times W$. *The equation* $R \odot_\sigma X = T^{*Y}$ *is solvable and* $R_Y \odot_\sigma X = T_Y$ *is a* Y*-reduced equation of* $R \odot_\sigma X = T^{*Y}$, *i.e.,* $T_w(u) = T_w^{*Y}(u)$ *for all* $(u, w) \in Y \times W$

Thus, this last result provides a first way to approximate a MARE.

Definition 9. *Let* $R \odot_\sigma X = T$ *be an unsolvable MARE and* Y *be a feasible reduct of its associated context* (U, V, R, σ). *We say that* $R \odot_\sigma X = T^{*Y}$ *is the* Y*-reduced approximation of the equation* $R \odot_\sigma X = T$ *if* $T_w^{*Y} = (T_Y)_w^{\downarrow_Y^N \uparrow_\pi}$ *for all* $w \in W$.

Note that, each feasible reduct gives rise to a different approximation of the original equation. Thus, depending on whether some conditions (equations) are considered more important than others, we can select one reduct (approximation) with respect to the others. Moreover, it need to be emphasized that not all reducts need to be necessarily feasible, and in particular not all reducts in the associated context have to give rise to an approximation of the unsolvable equation.

The following two approximation procedures were originally introduced in [4], in which FRE are used in abduction reasoning in the framework of logic programming. In a general setting, they consist in replacing T_w by the intension of some concept greater or smaller than T_w, for all $w \in W$.

Given the lattice $(\mathcal{I}(\mathcal{M}_{\pi N}), \preceq)$ of concepts associated with the Eq. (6), we will consider two sets containing, for each $w \in W$, those elements of $\mathcal{I}(\mathcal{M}_{\pi N})$ that are less than T_w or greater than T_w, respectively

$$L_w = \{f \mid \langle f^{\downarrow^N}, f \rangle \in \mathcal{M}_{\pi N} \text{ and } f \preceq_1 T_w\} \tag{7}$$

$$G_w = \{f \mid \langle f^{\downarrow^N}, f \rangle \in \mathcal{M}_{\pi N} \text{ and } T_w \preceq_1 f\} \tag{8}$$

Note that, substituting T_w by an element of L_w or of G_w, for all $w \in W$, gives rise to a solvable MARE since, by definition, the elements of these sets are intensions of concepts of $\mathcal{M}_{\pi N}$. The following result shows that the set L_w has a maximum element.

Proposition 3. *Let* $R \odot_\sigma X = T$ *be an unsolvable MARE. For each* $w \in W$, *the set* L_w *defined on (7) has a maximum element, which is given by* $T_w^{\downarrow^N \uparrow_\pi}$.

In general, the sets of the form G_w defined in (8) have no minimum element and, in certain cases, they may even be empty [15]. Therefore, we will use their minimal elements as an approximation, when they exist.

Definition 10. *Let* $R \odot_\sigma X = T$ *be an unsolvable MARE.*

1. *We say that* $R \odot_\sigma X = T_*$ *is the* conservative approximation *of the equation* $R \odot_\sigma X = T$ *if* $(T_*)_w = T_w^{\downarrow^N \uparrow^\pi}$ *for all* $w \in W$.
2. *We say that* $R \odot_\sigma X = T^*$, *is an* optimistic approximation *of the equation* $R \odot_\sigma X = T$ *if* T_w^* *is a minimal element of the set* G_w *defined in (8) for all* $w \in W$.

The following result shows that the conservative and optimistic approximations are the most suitable approaches if the independent term is desired to be increased or decreased at the same time in all its components, respectively. This means that it is not possible to find another approximation closer to the original solution with such features.

Proposition 4. *Let* $R \odot_\sigma X = T$ *be an unsolvable MARE. The following properties are satisfied:*

1. *If* $R \odot_\sigma X = T_*$ *is the conservative approximation of* $R \odot_\sigma X = T$, *there is no* $T' \in L_1^{V \times W}$ *such that* $R \odot_\sigma X = T'$ *is solvable and, for some* $w \in W$, *it is satisfied that* $(T_*)_w \prec_1 T_w' \prec_1 T_w$.
2. *If* $R \odot_\sigma X = T^*$ *is an optimistic approximation of* $R \odot_\sigma X = T$, *there is no* $T' \in L_1^{V \times W}$ *such that* $R \odot_\sigma X = T'$ *is solvable and, for some* $w \in W$, *it is satisfied that* $T_w \prec_1 T_w' \prec_1 T_w^*$.

The conservative and the optimistic approximation mechanisms are easily applicable and allow different alternatives to recover the solvability of a given MARE. Moreover, they always exist unlike the reduced approximation mechanism. On the contrary, the reduced method has a useful control on the modified elements in the independent term (equations to be removed), which does not apply to the conservative and the optimistic approaches.

The following example shows the conservative, optimistic and reduced approximations of an unsolvable MARE, illustrating some of the mentioned characteristics of these methods.

Example 1. Let $U = \{u_1, u_2, u_3, u_4, u_5\}$, $V = \{v_1, v_2, v_3, v_4, v_5\}$ and $W = \{w\}$ be three sets, and consider fixed the property-oriented multi-adjoint framework

$$([0,1]_8, \leq, \overline{\&_P}, \overline{\nearrow^P}, \overline{\nwarrow_P}, \overline{\&_2}, \overline{\nearrow^2}, \overline{\nwarrow_2})$$

where $(\overline{\&_P}, \overline{\nearrow^P}, \overline{\nwarrow_P})$ is the discretization of the adjoint triple associated to the product t-norm and $(\overline{\&_2}, \overline{\nearrow^2}, \overline{\nwarrow_2})$ is the discretization of the adjoint triple related to the conjunction $\&\colon [0,1] \times [0,1] \to [0,1]$ given by $x \,\&\, y = x^2 y$, for all $x, y \in [0,1]$. In this setting, consider the MARE

$$R \odot_\sigma X = T \qquad\qquad (9)$$

where $\sigma\colon V \to \{1,2\}$ assigns v_1, v_2, v_5 to the first adjoint triple and v_3, v_4 to the second one, while the relations $R \in [0,1]_8^{U \times V}$ and $T \in [0,1]_8^{U \times W}$ are defined as

$$R = \begin{pmatrix} 0.25 & 0.75 & 0.375 & 0.125 & 0.375 \\ 0.125 & 0.375 & 0.125 & 0 & 0.25 \\ 0.5 & 0.75 & 0.5 & 0.375 & 0.375 \\ 0.25 & 0.75 & 0.375 & 0.125 & 0.375 \\ 0.5 & 0.625 & 0.5 & 0.375 & 0.125 \end{pmatrix}, \qquad T = \begin{pmatrix} 0.25 \\ 0.125 \\ 0.375 \\ 0.125 \\ 0.125 \end{pmatrix}$$

First and foremost, notice that

$$T^{\downarrow^N \uparrow_\pi} = \begin{pmatrix} 0.25 \\ 0.125 \\ 0.5 \\ 0.875 \\ 0.25 \end{pmatrix}^{\uparrow_\pi} = \begin{pmatrix} 0.125 \\ 0.125 \\ 0.125 \\ 0.125 \\ 0.125 \end{pmatrix} \neq T$$

Hence, by Proposition 2, we can assert that (9) is not solvable. Let us then apply the different approximation techniques shown in this section.

To begin with, the conservative approximation of (9) is the MARE

$$R \odot_\sigma X = T_* \tag{10}$$

where the relation T_* is given by

$$T_* = T^{\downarrow^N \uparrow_\pi} = \begin{pmatrix} 0.125 \\ 0.125 \\ 0.125 \\ 0.125 \\ 0.125 \end{pmatrix}$$

Notice that, the first and the third elements of the right-hand side of the conservative MARE (10) are smaller than those of the original MARE (9), whilst the rest of elements remain unchanged. The associated concept is highlighted in blue in Fig. 1.

In order to compute the optimistic approximation(s) of (9), we will make use of the concept lattice $(\mathcal{M}_{\pi N}, \preceq_{\pi N})$ associated with the MARE, which is illustrated in Fig. 1. It holds that the set G_w (defined in (8)) has two minimal elements, which entail two optimistic approximations of (9):

$$R \odot_\sigma X = T_1^* \tag{11}$$

$$R \odot_\sigma X = T_2^* \tag{12}$$

where

$$T_1^* = \begin{pmatrix} 0.25 \\ 0.125 \\ 0.375 \\ 0.25 \\ 0.375 \end{pmatrix} \qquad T_2^* = \begin{pmatrix} 0.375 \\ 0.25 \\ 0.375 \\ 0.375 \\ 0.125 \end{pmatrix}$$

In this case, one of the optimistic approximations (11) increases the value of the two last components of the right-hand side of (9), while in the second optimistic approximation (12), the first, the second and the fourth element of the right-hand side of (9) have risen. In both cases, all other elements of the independent term are equal to the original MARE. The corresponding concepts are highlighted in orange in Fig. 1.

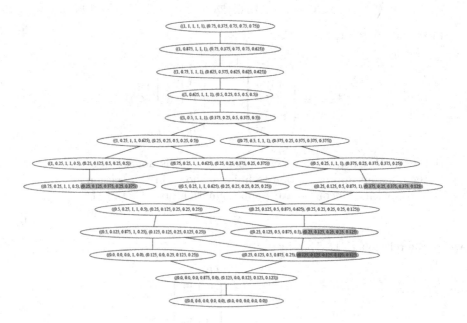

Fig. 1. Concept lattice associated with the MARE (9).

Lastly, making use of the procedures detailed in [8], it is satisfied that the context (U, V, R, σ) admits two reducts: $Y_1 = \{u_1, u_2, u_5\}$ and $Y_2 = \{u_2, u_4, u_5\}$. On the one hand, since

$$
T_{Y_1}^{\downarrow_{Y_1}^N \uparrow_\pi^{Y_1}} = \begin{pmatrix} 0.25 \\ 0.125 \\ 0.5 \\ 0.875 \\ 0.5 \end{pmatrix}^{\uparrow_\pi^{Y_1}} = \begin{pmatrix} 0.25 \\ 0.125 \\ 0.125 \end{pmatrix} = T_{Y_1}
$$

then the Y_1-reduced MARE

$$R_{Y_1} \odot_\sigma X = T_{Y_1}$$

is solvable. In other words, Y_1 is a feasible reduct and the right-hand side of the Y_1-reduced approximation of (9) is given by

$$T^{*Y_1} = T^{\downarrow_{Y_1}^N \uparrow_\pi} = \begin{pmatrix} 0.25 \\ 0.125 \\ 0.25 \\ 0.25 \\ 0.125 \end{pmatrix}$$

In this case, the third element of the right-hand side of (9) has been decreased and the fourth one has been increased, while the elements corresponding to the reduct Y_1 remain unaltered. Notice that, this compensational feature does not occur in the conservative or optimistic approximations (10), (11) and (12). This new approximation is associated with the concept highlighted in red in Fig. 1.

On the other hand, it can be checked that the reduct Y_2 is also feasible, but it does not lead to a new approximation, as the Y_2-reduced approximation of (9) coincides with the conservative approximation (10).

4 Selecting the Best Approximation

The three approximation mechanisms of unsolvable MARE presented in Sect. 3 are based on different philosophies for obtaining a new independent term such that the resulting MARE is solvable. Furthermore, two of these three mechanisms may give rise to more than one approximation, as shown in Example 1, where four different approximations of (9) have been calculated.

In order to contrast the approximations of an unsolvable MARE, a measure is necessary. In [19], three unsolvability measures were introduced to compare the original MARE with independent term T_w from its conservative approximation $T_w^{\downarrow^N \uparrow_\pi}$ [4]. Such study revealed how far is an unsolvable MARE from being solvable, assuming the (unique) conservative approximation. The idea here is defining measures applicable to all the possible approximations of a MARE, leading to a basis to compare the approximations between them. Notice that, for instance, a MARE can be far from its conservative approximation but close to one of its optimistic approximations.

It is clear that the first measure presented in [19] (which will be recalled next) perfectly works for the other two approximation approaches, while the other two measures are meaningless for optimistic and reduced approximations. In this section, we will consider the unit interval as the underlying lattice. Hence, we will use three distances between two columns/vectors for comparing the given independent term T and an approximation T^*. Given two columns C_1, C_2 with

k rows, we define the distance $\mathcal{D}_p \colon [0,1]^k \times [0,1]^k \to [0,1]$ as

$$\mathcal{D}_p(C_1, C_2) = \frac{\displaystyle\sum_{i=1}^{k} I(C_1[i], C_2[i])}{k}$$

for all $C_1, C_2 \in [0,1]^k$, where the mapping $I \colon [0,1] \times [0,1] \to \{0,1\}$ is defined as:

$$I(u,w) = \begin{cases} 0 & \text{if } u = w \\ 1 & \text{if } u \neq w \end{cases}$$

for all $u, w \in [0,1]$.

Basically, the measure $\mathcal{D}_p$ gives the average number of coordinates of the column C_1 that are different from the column C_2. Since this measure does not take into account the distance between the coordinates, another measure is required. Therefore, we will consider other natural distances, that is, the Manhattan, the Euclidean and the Chebyshev (or maximum metric) ones. Hence, the distances $\mathcal{D}_1, \mathcal{D}_2, \mathcal{D}_\infty$ defined for all $C_1, C_2 \in [0,1]^k$ as

$$\mathcal{D}_1(C_1, C_2) = \sum_{i=1}^{k} |C_1[i] - C_2[i]|$$

$$\mathcal{D}_2(C_1, C_2) = \sqrt{\sum_{i=1}^{k} (C_1[i] - C_2[i])^2}$$

$$\mathcal{D}_\infty(C_1, C_2) = \max\{|C_1[i] - C_2[i]| \mid i \in \{1, \ldots, k\}\}$$

will also be taken into account for comparing the columns of the original independent term of a MARE and the right-hand side of an approximation. Notice that, the resolution of a MARE of the form (6) is equivalent to the resolution of k subsystems

$$R \odot_\sigma X_w = T_w \tag{13}$$

where k is the number of columns of T and X_w, T_w are the w-th columns of X and T, respectively.

The following example applies the introduced measures to the approximations of (9) computed in Example 1.

Example 2. Continuing with Example 1, Table 1 shows the differences between the independent terms given by each approximate mechanism introduced above.

Therefore, in this example, the user will select the approximated FRE given by a reduct of the concept lattice associated with the FRE, which is the best approximation from the four possibilities. In the future, more relationships will be studied among the different obtained independent terms.

Table 1. Distances among the independent terms of the approximation mechanisms

Approximation Method	Independent term	$\mathcal{D}_p$	$\mathcal{D}_1$	$\mathcal{D}_2$	$\mathcal{D}_\infty$
Conservative	T_*	2/5	0.375	0.279508	0.25
Optimistic	T_1^*	2/5	0.375	0.279508	0.25
Optimistic	T_2^*	3/5	0.5	0.306186	0.25
Reduced	T^{*Y_1}	2/5	0.25	0.176777	0.125

5   Conclusions and Future Work

This paper has analyzed three mechanisms to approximate FRE based on the relationship between FRE and the two isotone variants of FCA. The first one takes advantage the attribute reduction results in FCA to remove problematic and redundant equations of the original FRE. As a consequence, unsolvable FRE becomes solvable because of removing possible inconsistencies in the systems of equations. The other two are based on looking for the intensions of the concept lattice associated with the given FRE that are closer to its independent term. Different measures have been proposed to analyze what approximation is the most appropriated in practical cases. They have also been applied to an example.

In the future, we will continue the study of approximations of unsolvable of FRE, introduce more measures and determine the impact of approximating FRE in their solution set.

References

1. Alcalde, C., Burusco, A., Díaz-Moreno, J.C., Medina, J.: Fuzzy concept lattices and fuzzy relation equations in the retrieval processing of images and signals. Int. J. Uncertain. Fuzziness Knowl.-Based Syst. **25**(Suppl. 1), 99–120 (2017)
2. Aliannezhadi, S., Abbasi Molai, A.: A new algorithm for geometric optimization with a single-term exponent constrained by bipolar fuzzy relation equations. Iran. J. Fuzzy Syst. **18**(1), 137–150 (2021)
3. Chen, J., Mi, J., Lin, Y.: A graph approach for knowledge reduction in formal contexts. Knowl.-Based Syst. **148**, 177–188 (2018)
4. Cornejo, M.E., Díaz-Moreno, J.C., Medina, J.: Multi-adjoint relation equations: a decision support system for fuzzy logic. Int. J. Intell. Syst. **32**(8), 778–800 (2017)
5. Cornejo, M.E., Lobo, D., Medina, J.: On the solvability of bipolar max-product fuzzy relation equations with the standard negation. Fuzzy Sets Syst. **410**, 1–18 (2021)
6. Cornejo, M.E., Lobo, D., Medina, J., De Baets, B.: Bipolar equations on complete distributive symmetric residuated lattices: the case of a join-irreducible right-hand side. Fuzzy Sets Syst. **442**, 92–108 (2022)
7. Cornejo, M.E., Medina, J., Ramírez-Poussa, E.: A comparative study of adjoint triples. Fuzzy Sets Syst. **211**, 1–14 (2013)
8. Cornejo, M.E., Medina, J., Ramírez-Poussa, E.: Characterizing reducts in multi-adjoint concept lattices. Inf. Sci. **422**, 364–376 (2018)

9. Cornejo, M.E., Medina, J., Ramírez-Poussa, E.: Algebraic structure and characterization of adjoint triples. Fuzzy Sets Syst. **425**, 117–139 (2021)
10. Cornelis, C., Medina, J., Verbiest, N.: Multi-adjoint fuzzy rough sets: definition, properties and attribute selection. Int. J. Approx. Reason. **55**, 412–426 (2014)
11. De Baets, B.: Analytical solution methods for fuzzy relation equations. In: Dubois, D., Prade, H. (eds.) The Handbooks of Fuzzy Sets Series, vol. 1, pp. 291–340. Kluwer, Dordrecht (1999)
12. Di Nola, A., Sanchez, E., Pedrycz, W., Sessa, S.: Fuzzy Relation Equations and Their Applications to Knowledge Engineering. Kluwer Academic Publishers, Norwell (1989)
13. Díaz-Moreno, J.C., Medina, J.: Multi-adjoint relation equations: definition, properties and solutions using concept lattices. Inf. Sci. **253**, 100–109 (2013)
14. Díaz-Moreno, J.C., Medina, J.: Solving systems of fuzzy relation equations by fuzzy property-oriented concepts. Inf. Sci. **222**, 405–412 (2013)
15. Díaz-Moreno, J.C., Medina, J.: Using concept lattice theory to obtain the set of solutions of multi-adjoint relation equations. Inf. Sci. **266**, 218–225 (2014)
16. Díaz-Moreno, J.C., Medina, J., Ojeda-Aciego, M.: On basic conditions to generate multi-adjoint concept lattices via Galois connections. Int. J. Gen. Syst. **43**(2), 149–161 (2014)
17. Gaume, B., Navarro, E., Prade, H.: A parallel between extended formal concept analysis and bipartite graphs analysis. In: Hüllermeier, E., Kruse, R., Hoffmann, F. (eds.) IPMU 2010. LNCS (LNAI), vol. 6178, pp. 270–280. Springer, Heidelberg (2010). https://doi.org/10.1007/978-3-642-14049-5_28
18. Kuznetsov, S.O.: Machine learning and formal concept analysis. In: Eklund, P. (ed.) ICFCA 2004. LNCS (LNAI), vol. 2961, pp. 287–312. Springer, Heidelberg (2004). https://doi.org/10.1007/978-3-540-24651-0_25
19. Lobo, D., López-Marchante, V., Medina, J.: On the measure of unsolvability of fuzzy relation equations. Stud. Comput. Intell. (2023, in press)
20. Lobo, D., López-Marchante, V., Medina, J.: Reducing fuzzy relation equations via concept lattices. Fuzzy Sets Syst. (2023)
21. Maio, C.D., Fenza, G., Gallo, M., Loia, V., Stanzione, C.: Toward reliable machine learning with congruity: a quality measure based on formal concept analysis. Neural Comput. Appl. **35**, 1899–1913 (2023)
22. Medina, J.: Multi-adjoint property-oriented and object-oriented concept lattices. Inf. Sci. **190**, 95–106 (2012)
23. Medina, J., Ojeda-Aciego, M., Valverde, A., Vojtáš, P.: Towards biresiduated multi-adjoint logic programming. In: Conejo, R., Urretavizcaya, M., Pérez-de-la-Cruz, J.-L. (eds.) CAEPIA/TTIA -2003. LNCS (LNAI), vol. 3040, pp. 608–617. Springer, Heidelberg (2004). https://doi.org/10.1007/978-3-540-25945-9_60
24. Pedrycz, W.: Fuzzy relational equations with generalized connectives and their applications. Fuzzy Sets Syst. **10**(1–3), 185–201 (1983)
25. Sanchez, E.: Resolution of composite fuzzy relation equations. Inf. Control **30**(1), 38–48 (1976)
26. Shao, M., Hu, Z., Wu, W., Liu, H.: Graph neural networks induced by concept lattices for classification. Int. J. Approx. Reason. **154**, 262–276 (2023)
27. Turunen, E.: On generalized fuzzy relation equations: necessary and sufficient conditions for the existence of solutions. Acta Universitatis Carolinae. Mathematica et Physica **028**(1), 33–37 (1987)
28. Valverde-Albacete, F., Peláez-Moreno, C.: Leveraging formal concept analysis to improve n-fold validation in multilabel classification. CEUR Workshop Proceedings, vol. 3151. CEUR-WS.org (2021)

Doubly-Lexical Order Supports Standardisation and Recursive Partitioning of Formal Context

Tim Pattison[✉] and Aryan Nataraja

Defence Science and Technology Group, West Avenue, Edinburgh, SA 5111, Australia
{tim.pattison,aryan.nataraja}@defence.gov.au

Abstract. Formal Concept Analysis (FCA) transforms a context bigraph, having vertices of type `object` and `attribute`, into a lattice digraph, whose vertices and arcs represent formal concepts and their covering relation. The computational complexity of most FCA algorithms is a polynomial function of the numbers of vertices in both the context bigraph and lattice digraph. While the latter quantity is fixed, the former can be decreased by context standardisation, a process which we show is facilitated by the efficient partition refinement algorithm of Spinrad.

The CARVE algorithm recursively partitions the context bigraph and corresponding lattice digraph by removing universal objects and attributes and their orphans and partitioning the resultant sub-context into its connected components. The associated software prototype uses the resultant tree structure to support coordinated browsing of both. This paper describes an additional, coordinated representation of the context bigraph which makes explicit in its bi-adjacency matrix the pattern of nested sub-contexts discovered by the CARVE algorithm. We show that permuting this matrix into doubly-lexical order with the aid of Spinrad's algorithm groups together into nested rectangles the bigraph edges belonging to these sub-contexts, and facilitates the two key processing steps of the CARVE algorithm.

1 Introduction

The CARVE algorithm and associated software prototype [7–9] recursively partition the context bigraph and corresponding lattice digraph of amenable [9] formal contexts, and use the resultant tree structure to support coordinated browsing of both. Each vertex in the tree corresponds to a sub-context bigraph and its associated lattice digraph, and hence also to a container in an inclusion tree layout of either graph [7,9]. Instead of automating the former inclusion tree layout for the current sub-context, however, the CARVE software prototype presents an unconstrained drawing of the bigraph. This paper describes an additional, coordinated representation of the sub-context bigraph which makes explicit in its bi-adjacency matrix the pattern of nested sub-contexts discovered by CARVE. This representation is a permutation of the rows and columns of the bi-adjacency

© Commonwealth of Australia 2023
D. Dürrschnabel and D. López Rodríguez (Eds.): ICFCA 2023, LNAI 13934, pp. 17–32, 2023.
https://doi.org/10.1007/978-3-031-35949-1_2

matrix into jointly reverse lectic (JRL) order, which reverses the doubly-lexical order [4,10]. We show that the JRL order groups together into nested rectangles the bigraph edges belonging to the nested sub-contexts identified by the CARVE algorithm. The process of efficiently calculating suitable row and column permutations has the beneficial side-effect that clarification and reduction of both objects and attributes become both simpler and more computationally efficient. Clarification and reduction – which are collectively referred to as context *standardisation* – reduce the size of a formal context without affecting the structure of the resultant concept lattice and, along with the reduced labelling scheme, diminish visual clutter when drawing the lattice directed acyclic graph (DAG). Standardising the formal context also avoids redundant computation on equivalent or derivative objects and attributes when enumerating the formal concepts of CARVE-indivisible sub-contexts.

The MAXMOD-PARTITION algorithm [1] uses partition refinement to partition the attributes of a formal context into equivalence classes known as *maxmods*. The attributes in a given maxmod have the same extents, and therefore appear as attribute labels on the same formal concept. Attribute clarification of the formal context involves choosing one attribute from each maxmod as its examplar, and removing any remaining attributes. Berry et al. [2] used the MAXMOD-PARTITION algorithm to prepare a formal context for attribute clarification. We show that the MAXMOD-PARTITION algorithm [1] produces a reverse lectic order of the attributes with respect to the prescribed order in which the objects were processed. This reverse lectic order is consistent with the partial order of attribute extents by set inclusion, such that an attribute precedes any others whose attribute extents are subsets. Reverse lectic order localises the computation of attribute closures by excluding subsequent columns of the context bi-adjacency matrix, thereby expediting computation of the arrow relations in preparation for attribute reduction of the formal context, and facilitates on-the-fly context reduction by ensuring correctness of the computed closures with respect to the attribute-reduced context. The MAXMOD-PARTITION algorithm [1] is readily adapted to find object equivalence classes in preparation for object clarification of the context, and to produce a reverse lectic order of objects in preparation for object reduction.

The reverse lectic orders of attributes and objects obtained by separate application of MAXMOD-PARTITION do not in general result in a *jointly* reverse lectic order of the context bi-adjacency matrix, even if partitioning of the objects is based on a reverse lectic order of the attributes and vice versa. Empirical observations suggest, however, that iterated alternation of MAXMOD-PARTITION over attributes and objects eventually converges to a jointly reverse lectic order. Alternatively, Spinrad's algorithm [10] efficiently permutes the rows and columns of a binary matrix into doubly-lexical order, with complexity linear in the number of possible edges in the context bigraph [10]. The rows and columns of a matrix in doubly-lexical order can be permuted into jointly reverse lectic (JRL) order simply by reversing both the row and column orders. The partition refinement approach used by Spinrad's algorithm simultaneously identifies the attribute and object equivalence classes needed for context clarification, as well as the reverse lectic orders of both the objects and attributes which facilitate context reduction.

The CARVE algorithm recursively removes fully-connected ("universal") objects and attributes and their orphaned neighbours and partitions the resultant sub-context into its connected components. We show that permuting the context bi-adjacency matrix into JRL order would facilitate these steps, and that JRL order would be preserved throughout the sub-contexts identified by the CARVE algorithm. For a clarified context in particular, only the first row of the bi-adjacency matrix need be tested for a universal object, and if one is identified, only the last column need be tested for an attribute orphaned by its removal. Similarly, only the first column and conditionally the last row need be tested in relation to a universal attribute. A breadth-first search for connected components could then be commenced from the edge at top left in the sub-context bi-adjacency matrix obtained by removing any universals and their orphans, and, after halting to remove the identified connected component from the sub-context bi-adjacency matrix, resumed from the top left of any remainder. This procedure should be iterated until there is no remainder, producing the sub-contexts in JRL order for subsequent analysis. In particular, the top-left to bottom-right order of the nested containers in the bi-adjacency matrix would then correspond to the left to right order of containers in the CARVE view of the lattice digraph.

2   Jointly Reverse Lectic Order

2.1   Lectic and Reverse Lectic Orders

Let the numbers $1, 2, .., n$ represent the ordinal positions of the members of an arbitrary ordered set, and let $\setminus$ denote set difference. For notational convenience, the lectic or lexical order over subsets of this set can be defined in terms of these ordinal positions as follows [5].

Definition 1. *Let $\mathcal{X}, \mathcal{Y} \subseteq \{1, 2, .., n\}$. Then $\mathcal{X} < \mathcal{Y}$ if there exists $i \in \mathcal{Y} \setminus \mathcal{X}$ such that $\mathcal{X} \cap \{1, 2, .., i-1\} = \mathcal{Y} \cap \{1, 2, .., i-1\}$*

Given some total order on the objects [attributes][1] of a formal context, two attribute extents [object intents] are in lectic order if and only if – henceforth *iff* – the first object [attribute] by which they differ belongs to the second attribute extent [object intent]. The lectic order over attribute extents [object intents] induces a total order over the attributes [objects]. Thus a prescribed total order over the objects [attributes] induces a lectic order over the attribute extents [object intents], which in turn induces a total "lectic" order over the attributes [objects]. The lectic order is unique up to attribute [object] equivalence.

Since $\mathcal{X} \subseteq \mathcal{Y} \Rightarrow \mathcal{X} \leq \mathcal{Y}$ [5], lectic ordering of the attribute extents [object intents] constrains the search for sets $\mathcal{X} \subseteq \mathcal{Y}$ to sets $\mathcal{X} \leq \mathcal{Y}$. This observation

---

[1] The truth of a sentence containing terms in square brackets is unchanged by substituting these terms for those which precede them.

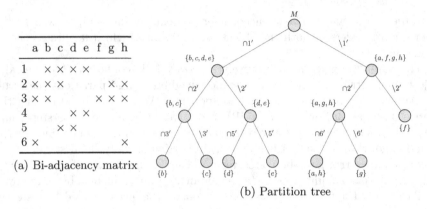

Fig. 1. Bi-adjacency matrix and attribute partition tree for context from [1].

has computational benefits for computing the arrow relations [6] in preparation for context reduction, as well as enumerating and transitively reducing the partial order between formal concepts. To this end, we reverse the lectic order, so that the lowest-ordered object [attribute] over whose membership two attribute extents [object intents] differ belongs to the set which *precedes* the other in the reverse lectic order. Since a set succeeds all of its supersets in the reverse lectic order, processing extents [intents] in reverse lectic order ensures that the extents of all super-concepts [intents of all sub-concepts] of a concept of interest are processed before it.

2.2 MAXMOD-PARTITION **and Reverse Lectic Order**

The MAXMOD-PARTITION algorithm for attribute clarification of a formal context produces a total order $<$ on the discovered attribute equivalence classes, and hence also on their exemplar attributes. Its iterated comparison of the next object intent with each part of the evolving attribute partition, and replacement of the latter with the intersection and remainder sets whenever both are non-empty, can be described by an ordered binary partition tree. Each left branch of this tree corresponds to an intersection, each right branch to a remainder, each non-leaf node to a part of an intermediate partition which has been subsequently refined, and each leaf node to an attribute equivalence class. The resultant ordered partition of the attribute set into equivalence classes is reflected in the left-right order of the corresponding leaf nodes. Figure 1a shows the partition tree for the example context in Fig. 1b. Each binary entry of the context bi-adjacency matrix is conventionally represented as either "$\times$" for 1 or " " for 0.

Proposition 1. *Let* $x, y \in M$ *belong to different attribute equivalence classes for a formal context* (G, M, I)*. Then* $x < y$ *if and only if the corresponding leaf nodes* *of the partition tree have a least common ancestor and lie respectively on the left and right branches of the sub-tree rooted at that ancestor.*

Proposition 2. MAXMOD-PARTITION *produces a reverse lectic order on the attribute extents with respect to the prescribed total order on objects.*

Proof. The object intents are processed according to the prescribed total order on objects. Let $x, y \in M$ belong to different attribute equivalence classes. Then the object whose intent split their least common ancestor is the first object in the prescribed order by which their attribute extents differ.

Without explicitly identifying this order as reverse lectic, Berry et al. [1] observed that an equivalence class can only *dominate* those which precede it.

Definition 2. *Attribute m dominates attribute μ if $m' \subset \mu'$.*

Here, $m' \subseteq G$ denotes the extent of attribute $m \in M$, which is the set of objects adjacent to m in the context bigraph. We saw in Sect. 2.1 that their observation is a consequence of the reverse lectic order.

2.3 Doubly-Lexical Order

Definition 3. *A context bi-adjacency matrix is in doubly-lexical order (DLO) [4, 10] if the objects (rows) are in lectic order with respect to the reverse order of the attributes (columns) and vice versa.*

Since a visualisation of the context bi-adjacency matrix which requires the user to mentally reverse the row and column orders would be cognitively demanding, we prefer to instead retain the row and column orders and define the *jointly reverse lectic* order accordingly as follows.

2.4 Jointly Reverse Lectic Order

Definition 4. *A context bi-adjacency matrix is in jointly reverse lectic (JRL) order if the objects (rows) are in reverse lectic order with respect to the order of the attributes (columns) and vice versa.*

Example 1. The tables in Figs. 2a and 2b show the bi-adjacency matrix of the example formal context from [7] in JRL and doubly-lexical order, respectively.

Proposition 3. *A context bi-adjacency matrix in DLO can be permuted into JRL order by reversing the rows and columns.*

Proposition 4. *The JRL order, and hence also the DLO, is not in general unique.*

The table in Fig. 3a illustrates Proposition 4 by providing an alternative JRL ordering of the context bi-adjacency matrix in Fig. 2b. The cells in the pink, light-blue and green rectangles are encountered in top-left to bottom-right order in Fig. 3a, whereas the contents of the light-blue rectangle are at bottom right in Fig. 2b. Similarly, the salmon rectangle occurs before the grey rectangle within the pink rectangle of Fig. 3a, whereas the contents of the former occur after those of the latter in Fig. 2b.

	(a) Doubly-lexical order													(b) JRL order												

(a) Doubly-lexical order

	L	K	D	C	A	B	E	F	J	I	H	G
7	×											
3	×	×										
4			×									
2						×						
11					×	×						
9			×	×	×	×						
10												×
5								×				×
12							×	×				×
6									×	×		×
1											×	×
8										×	×	×

(b) JRL order

	G	H	I	J	F	E	B	A	C	D	K	L
8	×	×	×									
1	×	×										
6	×		×	×								
12	×				×	×						
5	×				×							
10	×											
9							×	×	×	×		
11							×	×				
2							×					
4										×		
3											×	×
7												×

Fig. 2. Bi-adjacency matrix for context from [7] in doubly lexical and corresponding jointly reverse lectic orders.

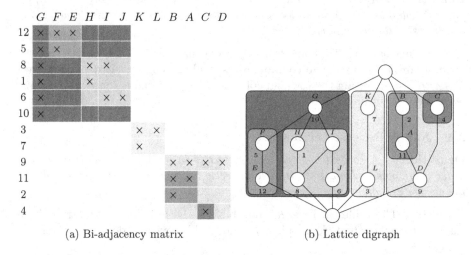

(a) Bi-adjacency matrix

(b) Lattice digraph

Fig. 3. Bi-adjacency matrix and lattice digraph for example formal context from [7]. Bi-adjacency matrix is in JRL order with cells colour-coded according to the inner-most corresponding CARVE container in the lattice digraph (Color figure online).

2.5   JRL Order and CARVE

Figure 3 illustrates that the nested containers in the inclusion layout of the lattice digraph in Fig. 3b are echoed as nested rectangles in the JRL-ordered bi-adjacency matrix. Each rectangle corresponds to a CARVE sub-context and contains the matrix entries corresponding to the internal edges of its bigraph. The left-to-right ordering of the containers in the lattice digraph corresponds to their top-left to bottom-right ordering in the JRL-ordered bi-adjacency matrix. Thus for this example, at least, permuting the bi-adjacency matrix of the context bigraph into JRL order groups together in contiguous and nested rectangles the edges between elements in the same CARVE sub-context. This anecdotal observation is formalised and generalised in Sect. 3.

A *universal* element of a context bigraph is one which is adjacent to all elements of the opposite type, and an element isolated by the removal of a universal element is referred to as its *orphan*. If a clarified formal context has a universal attribute [object] it must be first in the reverse lectic order, and its orphan object [attribute], if any, must be last. An element might only be "universal" with respect to its sub-context. For example, the pink sub-context of Fig. 3a has "universal" attribute G at left and orphan object 10 at bottom. Similarly, the green sub-context has "universal" object 9 at top and orphan attribute D at right. It is easily verified using Fig. 3a that removing these universals and their orphans from their respective sub-contexts leaves the remainder of each sub-context in JRL order, and with one or more connected components which are CARVE sub-contexts.

3   CARVE Identifies Nested Blocks in JRL-Ordered Matrix

3.1   Connected Components are Blocks

In this section, we show that permuting the bi-adjacency matrix into JRL order groups together in rectangles or "blocks" on the diagonal the objects and attributes belonging to each connected component of the context bigraph. We further describe a breadth-first search of the bigraph edges which identifies these blocks in the top-left to bottom-right order in which they appear in the JRL-ordered bi-adjacency matrix.

Let the context bi-adjacency matrix I be in JRL order, and label the rows so ordered with integers $1, 2, \ldots |G|$ and the columns so ordered with letters $A, B, \ldots$.

Lemma 1. *If $I(1, A) = 0$, then the context bigraph contains only isolated vertices.*

Proof. If $I(1, A) = 0$, then the first row of I must contain only zeros, since otherwise there is an attribute which belongs before A in the reverse lectic order of attributes. Similarly, the first column contains only zeros. The first column $j > A$ in any row $i > 1$ to contain a one would place row i before row 1 in the reverse lectic order, thereby contradicting the premise that I is in JRL order. Thus $I(i, j) = 0$ for all pairs (i, j).

Since a formal context consisting only of isolated vertices is fundamentally uninteresting, we henceforth safely assume that $I(1, A) = 1$, and hence also that object 1 is adjacent to attribute A. Consider now a breadth-first search (BFS) of the context bigraph edges – or equivalently the vertices of the corresponding line graph – starting at edge $(1, A)$. Depth 1 of this search discovers edges adjacent to edge $(1, A)$ – i.e. those adjacent to object 1 or attribute A – while excluding the edge $(1, A)$ "discovered" at depth 0. Depth 2 of this search discovers edges which are adjacent to those, but which were not discovered at depths 0 or 1, and so on. A vertex is said to be discovered at depth k of this search if it is adjacent to an edge discovered at depth k and is not adjacent to any previously-discovered edge.

To derive an expression for the set of vertices discovered at depth k, we define $\mathcal{N} : \mathcal{P}(G \cup M) \to \mathcal{P}(G \cup M)$ to map a set $\mathcal{X}$ of vertices to the union $\cup_{x \in \mathcal{X}} x'$ of their neighbours in the context bigraph, and denote by $\mathcal{N}^k = \mathcal{N} \circ \mathcal{N}^{k-1}$ the k-fold composition of this neighbourhood function. Here $\mathcal{N}^0$ is the identity function, and $\mathcal{N}(\emptyset) = \emptyset$. Note that since $\mathcal{N}(\{A\}) \supseteq \{1\}$, $\mathcal{N}^k(\{A\}) = \mathcal{N}^{k-1} \circ \mathcal{N}(\{A\}) \supseteq \mathcal{N}^{k-1}(\{1\})$ for $k \geq 1$ and similarly $\mathcal{N}^k(\{1\}) \supseteq \mathcal{N}^{k-1}(\{A\})$. For odd k

$$\mathcal{N}^k(\{A\}) = \left(\mathcal{N}^k(\{A\}) \setminus \mathcal{N}^{k-1}(\{1\})\right) \cup \left(\mathcal{N}^{k-1}(\{1\}) \setminus \mathcal{N}^{k-2}(\{A\})\right) \ldots \cup \{1\}$$
$$\mathcal{N}^k(\{1\}) = \left(\mathcal{N}^k(\{1\}) \setminus \mathcal{N}^{k-1}(\{A\})\right) \cup \left(\mathcal{N}^{k-1}(\{A\}) \setminus \mathcal{N}^{k-2}(\{1\})\right) \ldots \cup \{A\}$$

and for even k

$$\mathcal{N}^k(\{A\}) = \left(\mathcal{N}^k(\{A\}) \setminus \mathcal{N}^{k-1}(\{1\})\right) \cup \left(\mathcal{N}^{k-1}(\{1\}) \setminus \mathcal{N}^{k-2}(\{A\})\right) \ldots \cup \{A\}$$
$$\mathcal{N}^k(\{1\}) = \left(\mathcal{N}^k(\{1\}) \setminus \mathcal{N}^{k-1}(\{A\})\right) \cup \left(\mathcal{N}^{k-1}(\{A\}) \setminus \mathcal{N}^{k-2}(\{1\})\right) \ldots \cup \{1\}$$

The attribute and object sets discovered at depth 1 of the edge-based BFS are $\mathcal{N}(\{1\}) \setminus \{A\}$ and $\mathcal{N}(\{A\}) \setminus \{1\}$, respectively. At depth 2 they are $\mathcal{N}^2(\{A\}) \setminus \mathcal{N}(\{1\})$ and $\mathcal{N}^2(\{1\}) \setminus \mathcal{N}(\{A\})$, which, since attribute set $\mathcal{N}(\{1\}) = (\mathcal{N}(\{1\}) \setminus \{A\}) \cup \{A\}$ and object set $\mathcal{N}(\{A\}) = (\mathcal{N}(\{A\}) \setminus \{1\}) \cup \{1\}$, exclude vertices of the same type discovered at depths 1 and 0. Similarly at depth 3, attribute and object sets $\mathcal{N}^3(\{1\}) \setminus \mathcal{N}^2(\{A\})$ and $\mathcal{N}^3(\{A\}) \setminus \mathcal{N}^2(\{1\})$ respectively are discovered, with the second part of each expression excluding not only vertices of the same type discovered at depth 2, but also those at depths 1 and 0. Thus, for example, the set of objects discovered at odd depth k is

$$\mathcal{N}^k(\{A\}) \setminus \mathcal{N}^{k-1}(\{1\}) = \mathcal{N}^k(\{A\}) \setminus \Bigg(\left(\mathcal{N}^{k-1}(\{1\}) \setminus \mathcal{N}^{k-2}(\{A\})\right)$$
$$\cup \left(\mathcal{N}^{k-2}(\{A\}) \setminus \mathcal{N}^{k-3}(\{1\})\right) \ldots \cup \{1\} \Bigg)$$

The first excluded set on the right-hand side can be recognised as the union of the sets of objects discovered at depths $k-1, k-2 \ldots 0$. Similar results apply for even depths, and for attributes at even or odd depths.

Proposition 5. *A breadth-first search (BFS) of the edges of a context bigraph terminates when all and only objects and attributes in the connected component of the context bigraph containing the initial edge have been discovered.*

The order in which vertices are discovered during an edge-based BFS of the context bigraph starting at edge $(1, A)$ is related to the JRL order of the context bi-adjacency matrix as follows.

Theorem 1. *Let the bi-adjacency matrix of a formal context be in JRL order. For odd k, the attributes in $\mathcal{N}^k(\{1\}) \setminus \mathcal{N}^{k-1}(\{A\})$ are ordered after those in $\mathcal{N}^{k-1}(\{A\})$ and before those in $M \setminus \mathcal{N}^k(\{1\})$, and the objects in $\mathcal{N}^k(\{A\}) \setminus \mathcal{N}^{k-1}(\{1\})$ are ordered after those in $\mathcal{N}^{k-1}(\{1\})$ but before those in $G \setminus \mathcal{N}^k(\{A\})$. For even k, the attributes in $\mathcal{N}^k(\{A\}) \setminus \mathcal{N}^{k-1}(\{1\})$ are ordered after those in $\mathcal{N}^{k-1}(\{1\})$ but before those in $M \setminus \mathcal{N}^k(\{A\})$, and the objects in $\mathcal{N}^k(\{1\}) \setminus \mathcal{N}^{k-1}(\{A\})$ are ordered after those in $\mathcal{N}^{k-1}(\{A\})$ and before those in $G \setminus \mathcal{N}^k(\{1\})$.*

Proof. To prove Theorem 1 by induction on k, we first prove the case $k = 1$. Object 1 and attribute A are by definition first in their respective reverse lectic orders. Object 1 must be followed immediately by the remaining set of objects which have a one in their first column, and hence which are neighbours of attribute A, and then by the remaining objects – viz. those which are not neighbours of A. Similarly, attribute A is followed by the remaining neighbours of object 1 and thereafter by those attributes which are not.

We now show that if Theorem 1 holds for some $k \geq 1$, then it also holds for $k + 1$. In particular, we prove this for attributes in the case of even k; proofs for the remaining three parameter combinations are analagous. By our premise, the attributes are partitioned into three successive parts in the JRL order – viz. $\mathcal{N}^{k-1}(\{1\})$, $\mathcal{N}^k(\{A\}) \setminus \mathcal{N}^{k-1}(\{1\})$ and $M \setminus \mathcal{N}^k(\{A\})$. The union of the first two parts constitutes the first part for the case $k + 1$, while the third is partitioned into $\mathcal{N}^{k+1}(\{1\}) \setminus \mathcal{N}^k(\{A\})$ and $M \setminus \mathcal{N}^{k+1}(\{1\})$. Since the first two parts precede the third in the JRL order for case k, their union must precede its parts for case $k + 1$. It remains to prove that part $M \setminus \mathcal{N}^{k+1}(\{1\})$ follows part $\mathcal{N}^{k+1}(\{1\}) \setminus \mathcal{N}^k(\{A\})$ in the JRL order.

By our premise, the objects are similarly partitioned into three successive parts in the JRL order – viz. $\mathcal{N}^{k-1}(\{A\})$, $\mathcal{N}^k(\{1\}) \setminus \mathcal{N}^{k-1}(\{A\})$ and $M \setminus \mathcal{N}^k(\{1\})$. Whereas each attribute in $\mathcal{N}^k(\{A\}) = \mathcal{N} \circ \mathcal{N}^{k-1}(\{A\})$ is a neighbour of at least one object in $\mathcal{N}^{k-1}(\{A\})$, the columns corresponding to those in $M \setminus \mathcal{N}^k(\{A\})$ must have leading zeros in the rows corresponding to the objects in $\mathcal{N}^{k-1}(\{A\})$. Thus, the block of the bi-adjacency matrix corresponding to the first row part and the last column part for case k must contain only zeros.

Each attribute in $\mathcal{N}^{k+1}(\{1\}) \setminus \mathcal{N}^k(\{A\})$ is a neighbour of at least one object in $\mathcal{N}^k(\{1\}) \setminus \mathcal{N}^{k-1}(\{A\})$. Thus each column in $\mathcal{N}^{k+1}(\{1\}) \setminus \mathcal{N}^k(\{A\})$ has a 1 in at least one row of the second row part of case k, and hence in the first row part for case $k + 1$. In contrast, no attribute in $M \setminus \mathcal{N}^{k+1}(\{1\})$ is a neighbour of any object in the first row part $\mathcal{N}^k(\{1\})$ for case $k + 1$. Accordingly, the leading ones in the columns of $\mathcal{N}^{k+1}(\{1\}) \setminus \mathcal{N}^k(\{A\})$ occur earlier in the JRL order of rows than those of $M \setminus \mathcal{N}^{k+1}(\{1\})$, and hence column part $\mathcal{N}^{k+1}(\{1\}) \setminus \mathcal{N}^k(\{A\})$ precedes column part $M \setminus \mathcal{N}^{k+1}(\{1\})$ in the JRL order.

Corollary 1. *The objects and attributes discovered by a BFS from edge* $(1, A)$ *precede the remainder in the JRL order, and hence form a block of contiguous rows and columns at the top left of the context bi-adjacency matrix.*

Lemma 2. *The JRL order of this block is preserved by its removal from the JRL-ordered bi-adjacency matrix.*

Proof. Since no vertex in the connected component is adjacent to one which is not, any two vertices within the component can only differ with respect to their adjacency to other vertices in the component.

Lemma 3. *Removal of the initial connected component from the JRL-ordered context bi-adjacency matrix leaves what remains of the latter in JRL order.*

Proof. The remaining objects and attributes had no neighbours in the removed component, and hence their relative lectic order remains entirely dependent on their adjacency or otherwise to the remaining elements.

Lemma 4. *Permuting the bi-adjacency matrix* I *into JRL order groups together in contiguous blocks on the diagonal the objects and attributes belonging to each connected component of the context bigraph which contains at least one object and one attribute.*

Proof. Iterative application of Lemma 1, Corollary 1 and Lemma 3 ensures that for a context bigraph having multiple connected components, its JRL-ordered bi-adjacency matrix consists of one block on the diagonal per non-trivial connected component. Each isolated vertex of the formal context is a trivial "connected" component, for which the corresponding row or column is ordered after those for all non-trivial connected components.

If a formal context contains both isolated attributes and isolated objects, then these can be collectively viewed as a zero block which occurs last on the diagonal of I. If the only isolates are attributes, say, then the last block on the "diagonal" will end on the bottom row but to the left of the right-most column. If the context is clarified, then there is at most one isolate of each type.

Corollary 2. *A formal context whose bi-adjacency matrix is in JRL order can be partitioned into its connected components using a BFS of its edges which starts at* $(1, A)$ *and iteratively restarts at the top left entry of the bi-adjacency matrix of any remaining sub-context.*

When the search halts, the identified block of the bi-adjacency matrix is removed, and the BFS restarted at the top left element of the bi-adjacency matrix of any remaining sub-context. This procedure is iterated until no edges of the context bigraph remain undiscovered, in which case any remaining objects or attributes are isolates. Termination of the search can happen in any of four ways: both attributes and objects remain to be discovered but the top left element of the remaining sub-context is zero; or some objects and no attributes remain to be discovered; or some attributes and no objects remain to be discovered; or neither attributes nor objects remain to be discovered. A more conventional BFS of the *vertices* – vice edges – of the sub-context starting at either object 1 or attribute A is sufficient to identify the connected components in their JRL order.

3.2 Blocks are Nested

The CARVE algorithm recursively alternates between removing universals and their orphans and partitioning the resultant sub-context into its connected components. In this section, we show that it preserves clarification of the identified sub-contexts and the JRL order of their bi-adjacency matrices. When combined with Lemma 4, the latter result ensures that the connected components of the sub-context are contiguous blocks nested within that of the parent (sub-)context. We further show that both clarification and the JRL order afford efficiencies in the implementation of CARVE by significantly constraining which elements are candidates for universals, their orphans or isolates.

Proposition 6. *Removal by* CARVE *of any universal attributes and objects, along with any objects and attributes orphaned by their removal, does not affect the JRL of the bi-adjacency matrix of the remaining sub-context.*

Proof. Following the removal of any universal attributes, any difference between the objects continues to depend only on the remaining attributes. If the universal attributes had orphan objects, then any difference between the remaining attributes continues to depend only on the remaining objects. The same argument applies for universal objects and orphan attributes.

Only at the outermost level of the recursive CARVE algorithm, failure of the test for universal objects [attributes] is followed by a test for, and removal of any, isolated attributes [objects]. Since none of the objects [attributes] differ regarding their adjacency to an isolated attribute [object], any difference between them continues to depend only on the remaining attributes [objects]. Thus the JRL order remains unaffected.

Lemma 5. *The* CARVE *algorithm preserves the JRL order of the sub-context bi-adjacency matrices.*

Proof. Proposition 6 and the ensuing discussion demonstrates that removal by the CARVE algorithm of universals and their orphans, and, at the outermost level, any isolates, does not affect the JRL order of what remains of the context bi-adjacency matrix. Lemmas 2 and 3 establish that partitioning the remaining context into its connected components preserves the JRL order of the bi-adjacency matrices for those sub-contexts.

Theorem 2. *The connected components of the sub-context are contiguous blocks nested within that of the parent (sub-)context.*

Proof. This is a consequence of Lemmas 5 and 4.

The ordered partitions of the objects and attributes computed by Spinrad's algorithm [10] not only facilitate permutation of the context bi-adjacency matrix into JRL order, but also clarification of the context using the identified equivalence classes.

Proposition 7. *In a clarified formal context whose bi-adjacency matrix is in JRL order, only the first object and attribute need be tested for universality, and only the last object and attribute need be (conditionally) tested for isolates or the orphans of universals.*

Proposition 8. *The* CARVE *algorithm preserves sub-context clarification.*

Proof. Consider two objects in a clarified context which are neither universals, their orphans nor isolates, and hence which survive the removal of such objects. By our premise that the context is clarified, their intents must differ. Since they do not differ with respect to the membership of any universal, orphaned or isolated attributes, however, they continue to differ after removal of those attributes. These object intents also continue to differ after the resultant context bigraph is partitioned into its connected components, since this process neither adds nor removes neighbours.

Lemma 5 and Proposition 8 ensure that the shortcut in Proposition 7, which applies prior to recursion of the CARVE algorithm, continues to apply at all levels of the recursion.

4   Coordinated Browsing

The nested containers discovered by CARVE in the context bigraph correspond to nested containers in the lattice digraph. We have seen that the former containers also correspond to nested, contiguous blocks in the JRL-ordered context bi-adjacency matrix. A visualisation of the bi-adjacency matrix can be coordinated with these other views using the tree discovered by CARVE during its hierarchical decomposition of the context bigraph. As illustrated in Fig. 3a, each block can serve as a container whose appearance, including colour, should be chosen to emphasise its correspondence with its counterpart in the lattice digraph. The edge-based BFS described in Sect. 3.1 further ensures that as per Fig. 3, the top-left to bottom-right ordering of blocks consistently corresponds to the left-right order of the containers in the lattice digraph. Basing each of the coordinated views on the standardised context reduces visual clutter by reducing the number of visual elements which must be displayed. Nevertheless, the user should also be given the option to switch between the standardised and original contexts, with the latter requiring timely restoration of the removed elements, if any, to each of the views.

Visualisation of the JRL-ordered matrix improves on the containerised variant of a two-layered drawing of the context bigraph depicted in [9, Figure 5]. Edge crossings in the latter compromise its intelligibility for larger graphs. Without elision, a matrix view comfortably scales to sub-contexts having 100 or so objects and attributes. For larger contexts, eliding the content of subordinate containers may be necessary for the outermost containers, and the legibility of row and column labels will require careful attention.

5 Context Standardisation and Restoration

Spinrad's algorithm [10] identifies the object and attribute equivalence classes required for context clarification and facilitates permutation of the context bi-adjacency matrix into JRL order. Clarification chooses from each of the identified equivalence classes an exemplar, and deletes the remaining elements from the formal context. Those deleted elements can be restored to the intents and extents of the resultant concept lattice, and to the vertex labels of the corresponding lattice digraph, by substituting for each exemplar the corresponding equivalence class. As we saw in Sect. 2.1, reverse lectic orders of both the object intents and attribute extents also facilitates context reduction by ensuring that supersets occur earlier in the order than their subsets. As for clarification, reducible objects and attributes are removed from the context prior to enumeration of its formal concepts. Whereas the examplars serve as placeholders for the restoration of clarified context elements, no such placeholders exist for reduced elements. Nevertheless, the closure of each such element in the reduced context can serve as both the fingerprint and address of the lattice element to which it should be restored.

In this section, we show how these closures can be calculated on the fly during reduction of a clarified context in JRL order, and stored for subsequent use in the restoration phase. We further show that JRL order is unaffected by the deletion of reducible context elements, and hence that the sequential application of object and attribute reduction, in either order, not only correctly reduces the context, but also preserves its JRL order.

Algorithm 1. Efficient attribute reduction of formal context.

Require: M is ordered set of attributes in reverse lectic order
Require: M, I are attribute clarified
Ensure: M_r, I_r are attribute reduced and remain in reverse lectic order
Ensure: $\mathcal{H}(m) \subseteq M_r$ is reduced intent of attribute concept for reduced $m \notin M_r$

```
 1: function REDUCE-ATTRIBUTES(M,I)
 2:     M_r ← M
 3:     I_r ← I
 4:     for all m ∈ M do
 5:         Compute (m', m'')
 6:         U ← m'' \ m                    ▷ attributes μ : (μ', μ'') > (m', m'')
 7:         K ← U'                         ▷ extent of meet of ancestor attribute concepts
 8:         if K \ m' == ∅ then
 9:             H(m) ← U
10:             M_r ← M_r \ m                                              ▷ reduce
11:             I_r ← I_r[·, M_r]          ▷ delete m's column from context matrix
12:         end if
13:     end for
14:     return M_r, H, I_r
15: end function
```

Based on computation of the arrow relations [6], Algorithm 1 illustrates the process for attribute reduction; its extension to object reduction is straightforward. The correctness of Algorithm 1 relies on the reverse lectic order of the attributes to ensure that any reducible attributes other than m which are in the closure m'' with respect to the un-reduced context have been removed from the context before attribute m is processed. Computation of the reduced closure m'' involves only attributes $\mu \leq m$ in the reverse lectic order – since $\mu' \supseteq m' \Rightarrow \mu \leq m$ – and objects $g \in m'$. If the context were not clarified, the equivalence class of attribute m should be excluded in line 6, rather than just m.

Theorem 3. *The JRL order of the context bi-adjacency matrix I is unaffected by the removal of a reducible element.*

Proof. We prove the case where reducible attribute m is deleted, along with the corresponding column of I, and leave as an exercise for the reader the case where the reducible element is an object. Deletion of m does not affect the reverse lectic order of the remaining attributes because the object set and its total order are unchanged. Any pair of objects whose ordered intents, prior to the deletion, first differ by some attribute other than m, will remain in the correct relative order. Assume there exist $g, \gamma \in G$ whose ordered intents g' and γ' agree on the membership of attributes $\mu < m$, but disagree over the membership of the reducible attribute $m \in (m'' \setminus m)''$. Without loss of generality let $m \in g' \Longrightarrow m' \supseteq g'' \Longrightarrow m'' \subseteq g''' = g'$. Since the attributes in $m'' \setminus m$ are amongst those $\mu < m$ on which γ' and g' agree, $\gamma' \supseteq m'' \setminus m$. Hence the object concept for γ has intent $\gamma' = \gamma''' \supseteq (m'' \setminus m)'' \supseteq \{m\}$, which contradicts the premise that γ' disagrees with g' on the membership of m.

Example 2. Figure 4 shows the example context from Fig. 1a after JRL ordering followed by clarification and reduction. The reverse lectic order of the attributes differs from that discovered in Fig. 1b because the reverse lectic order on the objects differs from the total order shown in Fig. 1a. Here, attribute a is chosen as the exemplar for the equivalence class $\{a, h\}$ and attribute h is accordingly deleted. Reducible attribute g is then removed, and flagged for restoration to the formal concept having intent $\mathcal{H}(g) = \{a, b\}$ in the standardised context. These attributes are to the left of g in the reverse lectic order of attributes. As expected, the JRL order is preserved by both clarification and reduction. Had the context included a reducible attribute i, say, adjacent only to object 2, then attribute $g < i$ would have been removed by Algorithm 1 before i was processed, and hence excluded from its closure.

Any CARVE-indivisible sub-contexts of a standardised context are standardised. However, the CARVE algorithm relies on identifying and removing any universal – and hence also reducible – elements and their orphans from intermediate sub-contexts. While Proposition 6 ensures that JRL order is thereby maintained, we note in passing that the removal of any universal elements is also a special case of Theorem 3. Importantly, none of the results in Sect. 3 rely on the original context, or any intermediate sub-context, being reduced.

(a) Original

	a	h	b	g	c	f	d	e
2	×	×	×	×	×			
3	×	×	×	×		×		
6	×	×						
1			×		×		×	×
5					×		×	
4							×	×

(b) Clarified

	a	b	g	c	f	d	e
2	×	×	×	×			
3	×	×	×		×		
6	×						
1		×		×		×	×
5				×		×	
4						×	×

(c) Standardised

	a	b	c	f	d	e
2	×	×	×			
3	×	×		×		
6	×					
1		×	×		×	×
5			×		×	
4					×	×

Fig. 4. JRL-ordered bi-adjacency matrix for original, clarified and standardised context from [1].

6 Discussion

While both Spinrad's algorithm [10] and CARVE [9] hierarchically partition the objects and attributes of a formal context, only the latter contributes directly to transforming the context bigraph into the concept lattice digraph. Spinrad's partition refinement algorithm efficiently permutes the rows and columns of a binary matrix into doubly-lexical, and hence, trivially, jointly reverse lectic order. Its low computational complexity ($\mathcal{O}(|G||M|)$ [10]) is important because the context standardisation and CARVE decomposition it facilitates are both means to the end of identifying any CARVE-indivisible sub-contexts requiring analysis by a conventional FCA algorithm. CARVE's preservation of both JRL order and clarification further obviates repeated invocation of Spinrad's algorithm.

Brucker and Préa [4] used doubly-lexical ordering (DLO) of the context bi-adjacency matrix to superimpose on it the cover relation for *dismantlable* concept lattices. A lattice $\mathcal{L}$ is said to be dismantlable if it has an element x which is *doubly-irreducible* – i.e. has at most one upper and at most one lower cover – and $\mathcal{L} \setminus \{x\}$ is also dismantlable. The CARVE decomposition [9] instead targets only the supremum and infimum of the current sub-lattice, regardless of whether they are doubly-irreducible, and does not require $\mathcal{L}$ to be dismantlable. A concept lattice is dismantlable iff the associated context bigraph is chordal-bipartite [3]. A bipartite graph is chordal bipartite iff one can iteratively remove a *bi-simplicial* edge until no edge is left. An edge (x, y) is called bi-simplicial if $\mathcal{N}(\{x\}) \cup \mathcal{N}(\{y\})$ is a maximal biclique, and hence corresponds to a formal concept. Berry and Sigayret [3] used DLO of the bi-adjacency matrix to efficiently identify a bi-simplicial edge which, in a reduced context, corresponds to a doubly-irreducible lattice element. However, the need to reduce the context upon removal of the bi-simplicial edge makes iteration of this technique inefficient when generating an elimination scheme for the doubly-irreducible elements of $\mathcal{L}$ [3].

Theorem 2 established that the JRL order groups together in nested blocks the objects and attributes belonging to each sub-context identified by the recursive CARVE algorithm. Clarification and reduction of the formal context are facilitated by, and preserve, the JRL order, which can therefore be beneficial even if the context is not amenable [9] to CARVE. Similarly, the identification of universals, their orphans and isolates is facilitated by, and preserves, the JRL order. A breadth-first search of the edges of a JRL-ordered sub-context, commencing at edge $(1, A)$, identifies its first connected component. Following removal of this component, each successive connected component is identified from the remaining sub-context by iteratively restarting the search at the edge, if any, between the object and attribute thereby promoted to first in the JRL order. This procedure identifies the CARVE sub-contexts in the top-left to bottom-right order in which they appear in the bi-adjacency matrix, thereby facilitating correspondence with the left-to-right order of the containers in a drawing of the lattice digraph. A visualisation of the bi-adjacency matrix with its nested containers (Fig. 3a) has been proposed, which can be coordinated with that of the lattice digraph (Fig. 3b) using the decomposition tree discovered by CARVE. In future work these results will be exploited through updates to the CARVE prototype.

References

1. Berry, A., Bordat, J.P., Sigayret, A.: A local approach to concept generation. Ann. Math. Artif. Intell. (2007). https://doi.org/10.1007/s10472-007-9063-4
2. Berry, A., Gutierrez, A., Huchard, M., Napoli, A., Sigayret, A.: Hermes: a simple and efficient algorithm for building the AOC-poset of a binary relation. Ann. Math. Artif. Intell. (2014). https://doi.org/10.1007/s10472-014-9418-6
3. Berry, A., Sigayret, A.: Dismantlable lattices in the mirror. In: Cellier, P., Distel, F., Ganter, B. (eds.) ICFCA 2013. LNCS (LNAI), vol. 7880, pp. 44–59. Springer, Heidelberg (2013). https://doi.org/10.1007/978-3-642-38317-5_3
4. Brucker, F., Préa, P.: Totally balanced formal context representation. In: Baixeries, J., Sacarea, C., Ojeda-Aciego, M. (eds.) ICFCA 2015. LNCS (LNAI), vol. 9113, pp. 169–182. Springer, Cham (2015). https://doi.org/10.1007/978-3-319-19545-2_11
5. Ganter, B.: Two basic algorithms in concept analysis. In: ICFCA 2010, pp. 312–340 (2010). https://doi.org/10.1007/978-3-030-21462-3
6. Ganter, B., Wille, R.: Formal Concept Analysis. Springer, Heidelberg (1999). https://doi.org/10.1007/978-3-642-59830-2
7. Pattison, T., Weber, D., Ceglar, A.: Enhancing layout and interaction in formal concept analysis. In: IEEE Pacific Visualization Symposium, pp. 248–252. IEEE (2014). https://doi.org/10.1109/PacificVis.2014.21
8. Pattison, T.: Interactive visualisation of formal concept lattices. In: Burton, J., Stapleton, G., Klein, K. (eds.) Joint Proceedings of 4th International Workshop Euler Diagrams and 1st International Workshop Graph Visualization in Practice, vol. 1244, pp. 78–89. CEUR-WS.org (2014)
9. Pattison, T., Ceglar, A., Weber, D.: Efficient formal concept analysis through recursive context partitioning. In: Ignatov, D.I., Nourine, L. (eds.) Proceedings of 2018 International Conference CLA, Olomouc, Czech Republic, vol. 2123, pp. 219–230. CEUR-WS.org (2018)
10. Spinrad, J.P.: Doubly lexical ordering of dense 0–1 matrices. Inf. Process. Lett. **45**(5), 229–235 (1993). https://doi.org/10.1016/0020-0190(93)90209-R

Graph-FCA Meets Pattern Structures

Sébastien Ferré(✉)

Univ Rennes, CNRS, Inria, IRISA, 35000 Rennes, France
ferre@irisa.fr

Abstract. A number of extensions have been proposed for Formal Concept Analysis (FCA). Among them, Pattern Structures (PS) bring complex descriptions on objects, as an extension to sets of binary attributes; while Graph-FCA brings n-ary relationships between objects, as well as n-ary concepts. We here introduce a novel extension named Graph-PS that combines the benefits of PS and Graph-FCA. In conceptual terms, Graph-PS can be seen as the *meet* of PS and Graph-FCA, seen as sub-concepts of FCA. We demonstrate how it can be applied to RDFS graphs, handling hierarchies of classes and properties, and patterns on literals such as numbers and dates.

1 Introduction

Formal Concept Analysis (FCA) [17] has been applied to many different tasks – such as information retrieval, recommendation, ontology engineering, or knowledge discovery – and in many application domains, e.g., social sciences, software engineering, bioinformatics or chemoinformatics, natural language processing [15]. The variety of those tasks and application domains early called for FCA extensions in order to handle complex data. Complex data includes non-binary attributes, concrete domains, heterogeneous data, uncertain data, and structured data.

The earlier extensions enable to use complex descriptions of objects in place of sets of binary attributes. Three similar extensions have been introduced at almost the same time: Generalized Formal Concept Analysis [6], Logical Concept Analysis [12], and Pattern Structures [16]. They enable to describe objects with valued attributes, intervals over numbers and dates, convex polygons, partitions, sequence patterns, tree patterns, or labelled graph patterns [15]. Other extensions address the uncertainty of object descriptions [22], notably Fuzzy FCA [5] where the incidence between an object and an attribute is a truth degree in $[0, 1]$ instead of a crisp Boolean value. Triadic Concept Analysis [21] adds conditions to the incidence between an object and an attribute, making the formal context a ternary relation instead of a binary relation. Polyadic Concept Analysis [24] generalizes this idea by allowing any number of dimensions for the context. Finally, a number of more recent extensions add relationships between

S. Ferré—This research is supported by ANR project SmartFCA (ANR-21-CE23-0023).

D. Dürrschnabel and D. López Rodríguez (Eds.): ICFCA 2023, LNAI 13934, pp. 33–48, 2023.
https://doi.org/10.1007/978-3-031-35949-1_3

objects, so that the concepts do not depend only on the individual descriptions of objects but also on relationship patterns over interconnected objects. Relational Concept Analysis [23] combines several classical FCA contexts and several binary relations to form concepts whose intents are similar to description logic class expressions [3], combining object attributes, binary relations, and quantifier operators. Relational structures [19] and Graph-FCA [8] add n-ary relationships between objects ($n \geq 1$), and form n-ary concepts, i.e. concepts whose intents are equivalent to conjunctive queries, and whose extents are equivalent to the results of such queries, i.e. sets of n-tuples of objects.

In this paper, we propose to merge two FCA extensions that are representative of the first and last categories above: Pattern Structures (PS) and Graph-FCA. The aim is to combine the benefits of the two categories of extensions, in short complex descriptions and relationships between objects. Logical Concept Analysis (LCA) could have been used in place of PS but we have chosen PS as it has been more widely adopted, and because it is better suited to the effective computation of concepts. In this paper we choose to merge PS with Graph-FCA but it would be perfectly relevant to do so with Relational Concept Analysis (RCA). We hope this work will encourage and facilitate the merge with RCA in a future work. The merge results in a new FCA extension called Graph-PS. It is an elegant extension in the sense that PS is a special case of Graph-PS, obtained by not using inter-object relationships; and Graph-FCA is a special case of Graph-PS, obtained by using sets of binary attributes as descriptions of individual objects and inter-object relationships. As a consequence, classical FCA is also a special case of Graph-PS. It therefore acts as an unifying FCA theory encompassing classical FCA and two mainstream FCA extensions.

The paper is structured as follows. Section 2 recalls the main definitions and results of Pattern Structures and Graph-FCA, as preliminaries. Section 3 defines Graph-PS as the extension of Graph-FCA with PS-like descriptions, and illustrates the different notions with a running example combining binary relationships, valued attributes and intervals. Section 4 describes its application to RDFS graphs by defining a custom set of descriptions and similarity operator. Section 5 concludes the paper, and draws some perspectives.

2   Preliminaries

In this section, we recall the main definitions of two extensions of Formal Concept Analysis (FCA): Pattern Structures (PS) [16] and Graph-FCA [11]. The former extends FCA attributes with complex descriptions and patterns. The latter extends FCA with n-ary relations between objects, and n-ary concepts.

2.1   Pattern Structures (PS)

A *pattern structure* is a triple $K = (O, (D, \sqcap), \delta)$ where O is a set of objects, $(D, \sqcap)$ is a meet-semi-lattice of *patterns*, and $\delta \in O \to D$ is a mapping taking each object to its *description*. The meet operator $d_1 \sqcap d_2$ represents the *similarity* between two patterns d_1 and d_2. It entails a partial ordering $\sqsubseteq$, called *subsumption*, defined for all patterns $d_1, d_2 \in D$ as $d_1 \sqsubseteq d_2 \iff d_1 \sqcap d_2 = d_1$. Conversely, the pattern $d_1 \sqcap d_2$ is the most specific pattern, according to $\sqsubseteq$, that subsumes patterns d_1 and d_2. The *extension* of a pattern d is defined as the set of objects whose description contains the pattern[1].

$$ext(d) := \{o \in O \mid d \sqsubseteq \delta(o)\}, \quad \text{for every } d \in D$$

The *intension* of a set of objects X is defined as the most specific pattern that subsumes the description of all objects in X, i.e. the similarity between all those descriptions.

$$int(X) := \prod_{o \in X} \delta(o), \quad \text{for every } X \subseteq O$$

The two derivation operators (ext, int) form a Galois connection between the two posets $(2^O, \subseteq)$ and $(D, \sqsubseteq)$. A *pattern concept* of a pattern structure $K = (O, (D, \sqcap), \delta)$ is a pair (X, d) with $X \subseteq O$ and $d \in D$ such that $X = ext(d)$ and $d = int(X)$. The component X of a pattern concept (X, d) is called the *extent*, and the component d is called the *intent*. Pattern concepts are partially ordered by $(X_1, d_1) \leq (X_2, d_2) \Leftrightarrow X_1 \subseteq X_2 \Leftrightarrow d_2 \sqsubseteq d_1$. This partial ordering forms a complete lattice called the *pattern concept lattice*.

Pattern structures were first applied to labeled graphs [16,20], e.g. to discover frequent patterns in molecular structures. They were then applied to various types of descriptions: e.g., numbers and intervals [18], partitions for characterizing functional dependencies [4], RDF graphs [1].

Classical FCA is the special case of PS when $D := 2^A$ for a given set of attributes A, and the similarity between two sets of attributes is their intersection: $(\sqcap) := (\cap)$.

2.2   Graph-FCA

We first need to introduce notations for tuples. Given a set X of elements, a *k-tuple* of elements is an ordered collection of k elements that is written $(x_1, \ldots, x_k)$. For the sake of concision, a tuple is often written as an overlined letter $\overline{x}$, whose element at position i can be written $\overline{x}[i]$, or simply x_i if there is no ambiguity. The set of all k-tuples over X is written X^k. The set of all tuples of any arity is written $X^* = \bigcup_{k \geq 0} X^k$. The latter includes the empty tuple $()$ for arity 0.

Graph-FCA extends a formal context into a *graph context*, defined as a triple $K = (O, A, I)$ where O is a set of objects, A is a set of attributes, and $I \subseteq O^* \times A$ is an incidence relation between *tuples of objects* $\overline{o} \in O^*$ and attributes $a \in A$. Objects are graph nodes, attributes are graph labels, and an incidence $((o_1, \ldots, o_k), a) \in I$ – also written $a(o_1, \ldots, o_k)$ – is an ordered hyper-edge between nodes $o_1, \ldots, o_k$, labeled with a.

---

[1] The original notation for the two PS derivation operators is $(.)^\square$. We use the notations $ext(.)$ and $int(.)$ because they are more explicit and also consistent with notations in other FCA extensions.

A *projected graph pattern of arity k* (*k*-PGP) is a pair $Q = (\overline{x}, P)$ where $\overline{x} = (x_1, \ldots, x_k) \in \mathcal{V}^k$ is a tuple of k *projected variables*, and $P \subseteq \mathcal{V}^* \times A$ is a *graph pattern*. Variables are here graph nodes, and pattern elements $((y_1, \ldots, y_k), a)$ – also written $a(y_1, \ldots, y_k)$ – are hyper-edges. The set of k-PGPs $\mathcal{Q}_k$ is equipped, for each arity k, with a subsumption operator $\sqsubseteq_q$ and a similarity operator $\sqcap_q$. The *description* of a tuple of objects $\overline{o}$ is the PGP $Q(\overline{o}) := (\overline{o}, I)$, which uses objects as variables.

The *extension* of a k-PGP Q is defined as the set of k-tuples of objects whose description contains the PGP.

$$ext(Q) := \{\overline{o} \in O^k \mid Q \sqsubseteq_q Q(\overline{o})\}, \quad \text{for every } Q \in \mathcal{Q}_k$$

The *intension* of a set of k-tuples of objects R is defined as the most specific PGP that subsumes the description of all tuples of objects in R, i.e. the PGP-intersection of all those descriptions.

$$int(R) := \bigcap_{\overline{o} \in R} Q(\overline{o}), \quad \text{for every } R \subseteq O^k$$

Those two derivation operators (*ext*, *int*) form a Galois connection between the two posets $(2^{O^k}, \subseteq)$ and $(\mathcal{Q}_k, \sqsubseteq_q)$. A *graph concept* of a graph context $K = (O, A, I)$ is a pair (R, Q) with $R \subseteq O^k$ and $Q \in \mathcal{Q}_k$, for some arity k, such that $R = ext(Q)$ and $Q =_q int(R)$. The component R of a graph concept (R, Q) is called the *extent*, and the component Q is called the *intent*. Graph concepts are partially ordered by $(R_1, Q_1) \leq (R_2, Q_2) \Leftrightarrow R_1 \subseteq R_2 \Leftrightarrow Q_2 \sqsubseteq_q Q_1$. This partial ordering forms a complete lattice called the *graph concept lattice*. There is a distinct lattice for each arity.

Graph-FCA has been applied to syntactic representations of texts, either for mining linguistic patterns [10] or for information extraction [2]. It has also been applied to knowledge graphs through the notion of *concepts of neighbors* for approximate query answering [9] and for link prediction [14].

Classical FCA is a special case of Graph-FCA when only 1-tuples (i.e., singletons) are used, for the incidence relation, the projected variables of PGPs, and hence for concepts. This implies that PGPs and concept intents have the shape $((x), (\{a_1(x), \ldots, a_p(x)\}))$, they use a single variable x, and hence they are equivalent to sets of attributes $\{a_1, \ldots, a_p\}$. Concept extents are then sets of singleton objects, which are equivalent to sets of objects.

3   Graph-PS: Extending Graph-FCA with Pattern Structures

For concision sake, we reuse the terms of Graph-FCA in Graph-PS as the graph structure remains, and only the description of nodes and edges are affected by the extension. In the following, we first define graph contexts as a common generalization of Graph-FCA contexts and pattern structures. Then, we define Projected Graph Patterns (PGP) and operations on them as they play the role of concept intents. Finally, we define graph concepts, their formation through a Galois connection, and their organization into a lattice.

3.1   Graph Context

Definition 1 (graph context). *A graph context is a triple* $K = (O, (D, \sqcap), \delta)$, *where* O *is a set of* objects, *D is a meet-semi-lattice of* descriptions, *and* $\delta \in O^* \rightarrow D$ *is a mapping taking each* tuple *of objects to its* description. *The meet operator* $\sqcap$ *on descriptions entails a smallest description* $\bot$ *(called the* empty description*), and a partial ordering* $\sqsubseteq$ *(called* subsumption*), defined for all descriptions* $c, d \in D$ *as* $c \sqsubseteq d \iff c \sqcap d = c$.

Compared to Graph-FCA, each hyperedge is mapped to one description instead of to zero, one or several attributes. Graph-FCA is therefore equivalent to the special case of Graph-PS where descriptions are sets of attributes. An hyperedge that is mapped to the empty description is considered as a non-existent relationship: $\delta((o_1, o_2)) = \bot$ means that there is no relation from o_1 to o_2; in other words, the pair (o_1, o_2) is nothing more than a pair of objects. This can be paralleled with the blank cells $((o, a) \notin I)$ in a classical formal context.

Compared to PS, descriptions can not only be attached to objects but also to tuples of objects, which enables to express relationships between objects. Those relationships are taken into account, in addition to PS-like descriptions, when forming concepts. Concepts are sets of objects that have similar descriptions, and that also have similar relationships to objects that have similar descriptions, and so on.

Example 1. As an example of graph context K_{ex}, we extend an example from Graph-FCA about the British royal family, introducing taxonomic relationships between attributes, and numeric intervals. The objects are people belonging to three generations.

$$O_{ex} := \{Charles, Diana, William, Harry, Kate, George, Charlotte\}$$

They are respectively abbreviated as C, D, W, H, K, G, A. Objects (people) are described by a pair made of their gender and their birth year:

$$\delta(C) = man : 1948, \quad \delta(D) = woman : 1961,$$
$$\delta(W) = man : 1982, \quad \delta(K) = woman : 1982, \quad \delta(H) = man : 1984,$$
$$\delta(G) = man : 2013, \quad \delta(A) = woman : 2015.$$

Pairs of objects are described whether the second is a parent of the first, and with the rank among siblings.

$$\delta(W, C) = parent : 1 \quad \delta(H, C) = parent : 2 \quad \text{(same for D in place of C)}$$
$$\delta(G, W) = parent : 1 \quad \delta(A, W) = parent : 2 \quad \text{(same for K in place of W)}$$

For instance, $\delta(W, C) = parent : 1$ tells that William is the first child of Charles. Any other tuple of objects has the empty description: $\delta(\overline{o}) = \bot$.

We want to take into account similarities between attributes and values. First, we define the similarity between genders, $man \sqcap woman = person$, saying that men and women have in common to be persons. Second, like in [18], we define the similarity between two numeric values as the smallest interval that contains the two values: e.g., $1 \sqcap 3 = [1, 3]$. This extends to intervals by using the convex hull of two intervals, considering a value v as equivalent to the interval $[v, v]$.

$$[u_1, v_1] \sqcap [u_2, v_2] = [min(u_1, u_2), max(v_1, v_2)]$$

However, to avoid intervals that are too large and hence meaningless, intervals $[u, v]$ s.t. $v - u > \epsilon$ are generalized into the symbol $*$ that represents the range of all possible values. The threshold ϵ depends on the type of values: $\epsilon = 20$ for birth years so that similarity means "in the same generation", and $\epsilon = 1$ for birth ranks. To summarize, the set of descriptions in our example is defined as:

$$D_{ex} := \{a : [u, v], a : * \mid a \in \{person, man, woman, parent\}, u \leq v \in \mathbb{Z}\} \cup \{\bot\},$$

with $a : v$ as a shorthand for $a : [v, v]$, and $\bot$ the empty description. The similarity operator $d = d_1 \sqcap d_2$ is defined as follows. If $d_1 = \bot$ or $d_2 = \bot$ then $d = \bot$, otherwise, $d_1 = a_1 : V_1$ and $d_2 = a_2 : V_2$. Then, if the attribute similarity $a_1 \sqcap a_2$ is defined as attribute a, then $d = a : V$ where $V = V_1 \sqcap V_2$ as defined above with the ϵ threshold depending on the attribute, else $d = \bot$. □

3.2 Projected Graph Patterns (PGP)

A *graph pattern* over a graph context shares the same structure as a graph context, with nodes and hyperedges labelled by descriptions, except that nodes are variables that range over the objects of the context.

Definition 2 (graph pattern). *Let $K = (O, (D, \sqcap), \delta)$ be a graph context. A graph pattern over K is a pair $P = (V, \delta_P)$, where $V \subseteq \mathcal{V}$ is a finite set of variables (nodes), and $\delta_P \in V^* \to D$ is a mapping taking each hyperedge to its description (hyperedge label).*

Compared to Graph-FCA, hyperedges are labelled by custom descriptions rather than by sets of attributes. An *embedding* of a graph pattern in a graph context is a mapping $\phi \in V \to O$ from pattern variables to context objects such that for each hyperedge $\overline{x}$, the pattern description of the edge subsumes the context description of the corresponding edge, i.e. $\delta_P(\overline{x}) \sqsubseteq \delta(\phi(\overline{x}))$.

Example 2. Given the graph context in Example 1, we introduce the example graph pattern $P_{ex} = (\{x, y\}, \delta_P)$, where the description δ_P is defined as follows:

$$\delta_P(x) = man : [1980, 1989], \quad \delta_P(y) = person : *, \quad \delta_P(x, y) = parent : [1, 2].$$

This description can be more concisely written as follows:

$$\delta_P = \{x \mapsto man : [1980, 1989], y \mapsto person : *, (x, y) \mapsto parent : [1, 2]\}.$$

This pattern represents the situation where a man born in the eighties (x) is the first or second child of some person with unconstrained birthdate (y). The pattern has four embeddings in the context, e.g. $\{x \mapsto Harry, y \mapsto Diana\}$ because Harry is a man born in 1984, and is the second child of Diana, who is a woman and hence a person. □

A *Projected Graph Pattern (PGP)* is a graph pattern with a tuple of distinguished variables, called *projected variables*.

Definition 3 (PGP). *A projected graph pattern (PGP) is a couple* $Q = (\overline{x}, P)$ *where* $P = (V, \delta)$ *is a graph pattern, and* $\overline{x} \in V^*$, *called* projection tuple, *is a tuple of variables from the pattern.* $|Q| = |\overline{x}|$ *denotes the* arity *of the PGP. We note* $\mathcal{Q}$ *the set of PGPs, and* $\mathcal{Q}_k$ *the subset of k-PGPs, i.e. PGPs having arity k.*

A PGP can be seen as a SPARQL query SELECT $\overline{x}$ FROM { P }, whose answers are the embeddings of the pattern restricted to the projected variables.

Example 3. The PGP $Q_{ex} = ((x), P_{ex})$ based on the graph pattern in Example 2 selects all men born in the eighties as the first or second child of somebody. The answers over the example context are therefore Harry and William. The PGP $((x, y), P_{ex})$ would select pairs *(child, parent)*, such as (Harry, Diana).   □

In Graph-FCA and Graph-PS, PGPs play the role of descriptions in PS. We therefore have to define two key operations on them: inclusion $\sqsubseteq_q$ (aka. subsumption) and intersection $\sqcap_q$ (aka. similarity). In Graph-PS, their definitions depend on the corresponding operations on PS-like descriptions, $\sqsubseteq$ and $\sqcap$.

Definition 4 (PGP inclusion). *Let* $K = (O, (D, \sqcap), \delta)$ *be a graph context. Let* $Q_1 = (\overline{x}_1, (V_1, \delta_1))$, $Q_2 = (\overline{x}_2, (V_2, \delta_2))$ *be two k-PGPs for some arity k.* Q_1 *is included in* Q_2, *or equivalently* Q_2 *contains* Q_1, *which is written* $Q_1 \sqsubseteq_q Q_2$ *iff*

$$\exists \phi \in V_1 \to V_2 : \phi(\overline{x}_1) = \overline{x}_2 \wedge \forall \overline{y} \in V_1^* : \delta_1(\overline{y}) \sqsubseteq \delta_2(\phi(\overline{y}))$$

According to this definition, the inclusion of Q_1 into Q_2 is analogous to the embedding of a pattern into a context, with the difference that variables are mapped to the variables of another pattern instead of the objects of the context. There is also the additional constraint that the projected variables match.

Example 4. For example, the PGP $Q' = ((z), (\{z\}, \{z \mapsto man:*\}))$, which selects the set of men, is included in the above PGP Q_{ex}, through the embedding $\phi = \{z \mapsto x\}$.   □

Definition 5 (PGP intersection). *Let* ψ *be an injective mapping from pairs of variables to fresh variables. The intersection of two k-PGPs* $Q_1 = (\overline{x}_1, (V_1, \delta_1))$ *and* $Q_2 = (\overline{x}_2, (V_2, \delta_2))$, *written* $Q_1 \sqcap_q Q_2$, *is defined as* $Q = (\overline{x}, (V, \delta))$, *where*

$$\overline{x} = \psi(\overline{x}_1, \overline{x}_2),$$
$$V = \{\psi(v_1, v_2) \mid v_1 \in V_1, v_2 \in V_2\},$$
$$\delta(\overline{y}) = \delta_1(\overline{y}_1) \sqcap \delta_2(\overline{y}_2), \text{ for } \overline{y} = \psi(\overline{y}_1, \overline{y}_2) \in V^*$$

PGP intersection works as a product of two PGPs where each pair of edges $(\overline{y}_1, \overline{y}_2)$ makes an edge whose description is the similarity $\delta_1(\overline{y}_1) \sqcap \delta_2(\overline{y}_2)$ between the descriptions of the two edges.

Example 5. The intersection of Q_{ex} and Q' results in the PGP

$$Q'' = ((xz), (\{xz, yz\}, \{xz \mapsto man : *, yz \mapsto person : *\})).$$

Variables xz and yz result from the pairing of variables from each PGP (function ψ). The tuples of variables that are not shown in the δ_P part have the empty description. For instance, $\delta_P((xz, yz)) = \delta_{ex}((x, y)) \sqcap \delta'((z, z)) = parent : [1, 2] \sqcap \perp = \perp$.

In Q'' the description of yz is disconnected from the projected variable xz, and is therefore useless to the semantics of Q''. Q'' can therefore be simplified to $((xz), (\{xz\}, \{xz \mapsto man : *\}))$, which is equal to Q' up to renaming variable xz as x. More information about such simplifications are available in [11].   □

The above example suggests as expected that $Q_1 \sqsubseteq_q Q_2$ implies $Q_1 \sqcap_q Q_2 = Q_1$. The following lemma proves that this is indeed the case, like with PS descriptions.

Lemma 1. *Let Q_1, Q_2 be two PGPs. Their PGP intersection $Q_1 \sqcap_q Q_2$ is their infimum relative to query inclusion $\sqsubseteq_q$.*

Proof. To prove that $Q = Q_1 \sqcap_q Q_2$ is a lower bound, it suffices to prove that Q is included in both Q_1 and Q_2. To prove $Q \sqsubseteq_q Q_1$, it suffices to choose the mapping $\phi_1(x) = (\psi^{-1}(x))[1]$ (recall that ψ is an injective mapping from 2-tuples of variables to variables), and to prove that $\phi_1(\bar{x}) = \bar{x}_1$ and $\delta(\bar{y}) \sqsubseteq \delta_1(\phi_1(\bar{y}))$ for all $\bar{y} \in V^*$. This is easily obtained from the definition of Q. The proof of $Q \sqsubseteq_q Q_2$ is identical with $\phi_2(x) = (\psi^{-1}(x))[2]$.

To prove that $Q_1 \sqcap_q Q_2$ is the *greatest* lower bound (the infimum), we have to prove that every PGP Q' that is included in both Q_1 (via ϕ_1) and Q_2 (via ϕ_2) is also included in Q. To that purpose, it suffices to choose $\phi(x') = \psi(\phi_1(x'), \phi_2(x'))$, and to prove that $\phi(\bar{x}') = \bar{x}$ and $\delta'(\bar{y}') \sqsubseteq \delta(\phi(\bar{y}'))$ for all $\bar{y}'$. This can be obtained from the definition of Q, and from the hypotheses.   □

3.3   Graph Concepts

As usual in FCA, concepts are composed of an extent and an intent. In Graph-PS like in Graph-FCA, k-PGPs in $\mathcal{Q}_k$ play the role of intents. For the extents we use the answers of PGPs seen as queries, i.e. sets of tuples of objects. The latter are mathematically k-ary *relations* over objects: $R \subseteq O^k$, for some arity $k \geq 0$. We note $\mathcal{R}_k = 2^{O^k}$ the set of k-relations over the objects of some graph context K.

Example 6. The set of father-mother-child triples can be represented as the following 3-relation (with abbreviated people names).

$$R := \{(C, D, W), (C, D, H), (W, K, G), (W, K, A)\}$$

Note that the order of objects in tuples matters while the order of tuples in the relation does not. A k-relation can be seen as a table with k unlabeled columns.

Charles	Diana	William
Charles	Diana	Harry
William	Kate	George
William	Kate	Charlotte

□

Before defining the Galois connection between PGPs and relations, we introduce the notion of *graph description* $\gamma(\bar{o})$ of an object or a tuple of objects. It incorporates everything that is known about an object or tuple of objects, in terms of relationships in the graph context around those objects, and in terms of D-description of those relationships. It therefore integrates the description δ of individual hyperedges.

Definition 6 (graph description). *Given a graph context* $K = (O, (D, \sqcap), \delta)$, *the* graph description *of any object* $o \in O$ *is defined as the PGP* $\gamma(o) := ((o), P_K)$ *where* $P_K = (O, \delta)$. *By extension, the description of any tuple of objects* $\bar{o} \in O^*$ *is defined as* $\gamma(\bar{o}) := (\bar{o}, P_K)$.

This definition says that the graph description of an object is the whole graph context, seen as a graph pattern (objects as variables), and projected on the object. In practice, only the part of the graph context that is connected to the object is relevant. The generalization to tuples of objects enables to have a description for pairs of objects, triples of objects, and so on.

From there, we can define two derivation operators between PGPs and relations, and prove that they form a Galois connection.

Definition 7 (extension). *Let* $K = (O, (D, \sqcap), \delta)$ *be a graph context. The* extension *of a* k-PGP $Q \in \mathcal{Q}_k$ *is the* k-relation defined by

$$ext(Q) := \{\bar{o} \in O^k \mid Q \subseteq_q \gamma(\bar{o})\}$$

The extension of a k-PGP is the set of k-tuples of objects whose graph description contains the PGP. It can be understood as the set of answers of the PGP seen as a query.

Example 7. In the example context, the extension of the 2-PGP

$$Q = ((x, y), (\{x, y\}, \{x \mapsto person : *, y \mapsto woman : *, (x, y) \mapsto parent : 1\}))$$

is the 2-relation

$$R = ext(Q) = \{(William, Diana), (George, Kate)\},$$

i.e. the set of pairs *(first child, mother)*.   □

Definition 8 (intension). *Let* $K = (O, A, I)$ *be a graph context. The* intension *of a* k-relation $R \in \mathcal{R}_k$ *is the* k-PGP defined by

$$int(R) := \bigcap_{\bar{o} \in R} \gamma(\bar{o})$$

The intension of a k-relation is the PGP intersection of the graph descriptions of all tuples of objects in the relation, hence the most specific projected graph pattern shared by them.

Example 8. In the example context, the intension of the 2-relation from Example 7

$$R = \{(William, Diana), (George, Kate)\}$$

is the 2-PGP

$$Q = int(R) = ((x, y), (\{x, y, z, w\}, \delta_P))$$

where $\delta_p = \{x \mapsto man\!:\!*,\ y \mapsto woman\!:\!*,\ z \mapsto man\!:\!*,\ w \mapsto person\!:\!*,$ $(x, y) \mapsto parent\!:\!1,\ (x, z) \mapsto parent\!:\!1,\ (w, y) \mapsto parent\!:\!2,\ (w, z) \mapsto parent\!:\!2\}$. Note that this intension expands the PGP in Example 7 with the following elements: x is a man, x is the first child of some man z, his father, and there is a second child w of parents y and z. The extension of this expanded PGP remains the relation R, which suggests that $int \circ ext$ is a closure operator. □

We can actually prove that ext and int form a Galois connection. This implies that $int \circ ext$ and $ext \circ int$ are closure operators, respectively on PGPs and relations.

Theorem 1 (Galois connection). *Let $K = (O, (D, \sqcap), \delta)$ be a graph context. For every arity k, the pair of mappings (ext, int) forms a Galois connection between $(\mathcal{R}_k, \subseteq)$ and $(\mathcal{Q}_k, \sqsubseteq_q)$, i.e. for every object relation $R \in \mathcal{R}_k$ and PGP $Q \in \mathcal{Q}_k$,*

$$R \subseteq ext(Q) \Longleftrightarrow Q \sqsubseteq_q int(R)$$

Proof. $R \subseteq ext(Q) \iff \forall \overline{o} \in R : \overline{o} \in ext(Q)$
$\iff \forall \overline{o} \in R : Q \sqsubseteq_q \gamma(\overline{o})$ (Definition 7)
$\iff Q \sqsubseteq_q \bigsqcap_{\overline{o} \in R} \gamma(\overline{o})$ (Lemma 1)
$\iff Q \sqsubseteq_q int(R)$ (Definition 8) □

From the Galois connection, *graph concepts* can be defined and organized into concept lattices, like in classical FCA, with one concept lattice for each arity k.

Definition 9 (graph concept). *Let $K = (O, (D, \sqcap), \delta)$ be a graph context. A k-graph concept of K is a pair (R, Q), made of a k-relation (the extent) and a k-PGP (the intent), such that $R = ext(Q)$ and $Q =_q int(R)$.*

Example 9. The 2-relation and 2-PGP in Example 8 form a 2-graph concept. It can be understood as the *(first child, mother)* binary relationship. Its intent tells us that in the example context, every first child whose mother is known also has a known father, and a sibling (man or woman) that was born after him. □

Theorem 2 (graph concept lattices). *The set of graph k-concepts $\mathcal{C}_k$, partially ordered by $\le$, which is defined by*

$$(R_1, Q_1) \le (R_2, Q_2) : \iff R_1 \subseteq R_2 \iff Q_2 \sqsubseteq_q Q_1,$$

forms a bounded lattice $(\mathcal{C}_k, \le, \wedge, \vee, \top, \bot)$, the k-graph concept lattice.

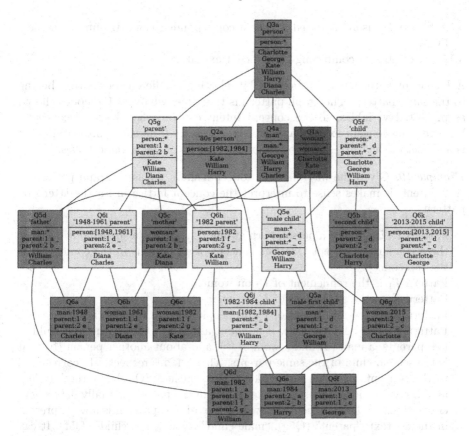

Fig. 1. The 1-graph concept lattice of the royal family context (less the bottom concept). The descriptor `parent:1 a _` in box Q5g reads $\delta_{P5}(a, g) = parent:1$, as part of the graph pattern $P5$ of the intent of concept Q5g. The nodes of pattern $P5$ are the boxes Q5a-g.

Figure 1 is a representation of the 1-graph concept lattice – less the bottom concept – of the royal family graph context from Example 1. This is a compact representation because each box represents at the same time a 1-concept and a node of the graph pattern of a concept intent. Each box is made of three parts:

1. A concept/node identifier Qnx made of a number n and a letter x. The number identifies a pattern P_n, and the letter identifies a node of this pattern. Together they form the concept intent $Q_n = (x, P_n)$. The set of nodes V_n of pattern $P_n = (V_n, \delta_{P_n})$ is therefore made of all boxes whose identifier has Qn as a prefix. (Some concepts also have an informal description under the identifier, manually added to help the reading of the lattice.)
2. A list of pattern hyperedges in the form $d\ y_1\ \dots\ y_n$, to be read as $\delta_{P_n}(y_1, \dots, y_n) = d$, where an underscore _ refers to the current node x.

A 1-edge $d\_$ is abbreviated as d as it corresponds to an attribute in classical FCA.

3. A list of objects composing the concept extent.

A k-concept with $k > 1$ can be read by selecting k different boxes that belong to the same pattern. The graph pattern is the same, whatever the nodes chosen as projected variables, so the concept intent can be read like for 1-concepts. However, the concept extent cannot be read exactly from this representation, it is a subset of the Cartesian product of the extents of the selected boxes.

Example 10. Given the way different concepts share the same graph pattern in their intent, it makes sense to interpret the concept lattice in Fig. 1 pattern by pattern.

- Pattern P_3 is a about a person born at any time. It has a single node a, hence there is a single concept Q3a, the concept of all persons in the context. It is the top concept.
- Pattern P_1 is the refinement of P_3 on women.
- Pattern P_4 is the refinement of P_3 on men.
- Pattern P_2 is the refinement of P_3 on people born in the eighties.
- Pattern P_5 has 7 nodes (a-g). It is about a man (a) that is the first child of two parents, a man (d) and a woman (c), and about another person (b) that is the second child of the same parents. Those nodes respectively correspond to the concepts of "male first child" (Q5a), "father" (Q5d), "mother" (Q5c), and "second child" (Q5b). The other nodes in the pattern actually define generalizations of those concepts for which the a-b-c-d pattern is always present in the context: "parent" (Q5g), "male child" (Q5e), and "child" (Q5f). It can be observed that the latter concepts (e-g) are in a lighter color and the former concepts (a-d) are in a more vivid color (called *core concepts/nodes* [11]). When reading a concept intent, the boxes in a lighter color can be ignored when they are not selected because the information they provide is redundant with the core nodes.
- The core nodes of pattern P_6 reproduce the graph context, each node corresponding to a specific object. The non-core concepts provide generalizations over single objects: "grand-parents born between 1948 and 1961" (Q6i), "parents born in 1982" (Q6h), "children born between 1982 and 1984" (Q6j), and "grand-children born between 2013 and 2015" (Q6k).

From the concept lattice structure, it is possible to find what different people have in common. For instance, from $Q6d \vee Q6g = Q5f$, we learn that what William (Q6d) and Charlotte (Q6g) have in common is that they are persons with a father (Q5d, Charles or William) and a mother (Q5c, Diana or Kate) who have together a male first child (Q5a, William or George), and a second child (Q5b, Harry or Charlotte). We also learn that they share that with Harry and George, the other instances of concept Q5f.   □

4 Application to RDFS Graphs

As an application case of Graph-PS we consider RDFS graphs. An RDFS graph is a structure $\langle R, \mathcal{L}, (\mathcal{C}, \leq), (\mathcal{P}, \leq), T \rangle$, where R is a collection of resources (IRIs and blank nodes), $\mathcal{L}$ is a set of literal values of various datatypes (e.g., strings, numbers, dates), $\mathcal{C}$ is a hierarchy of classes, $\mathcal{P}$ is a hierarchy of properties, and T is a set of triples expressing the factual knowledge.

In order to apply Graph-PS to RDFS graphs, we need to identify what are the objects, the descriptions, and the similarity between descriptions. From the usual RDFS graph representation that uses resources, literals, and classes as nodes, it is tempting to use them as objects. However, it is desirable to define similarity over literals and classes. The similarity between two integer literals could be an interval like in the example of the previous section. The similarity between two classes should be the most specific common ancestor class. Moreover, we think that literals and classes are more appropriate as descriptors of objects than as objects to be described. We therefore define the set of object as $O = R$.

We now look at the description of (tuples of) objects. The description information lies in the triples. We define below their conversion into elementary descriptions, according to the three kinds of triples.

- $(r, \texttt{rdf:type}, c) \rightsquigarrow \delta(r) = \{c, \ldots\}$
  The triple states that resource r is an instance of class c. The class is used as a descriptor of the resource, $\texttt{rdf:type}$ can be ignored because it is always used with a class. We use an open set containing c because a resource can be declared an instance of several classes.
- $(r, p, r') \rightsquigarrow \delta(r, r') = \{p, \ldots\}$
  The triple states that resource r is related to resource r' with property p. The property is used as a descriptor of the pair of resources, hence representing a binary edge. We again use an open set because RDFS graphs are multigraphs, i.e. several properties can relate the same resources (e.g., a person who is both the director and an actor of some film).
- $(r, p, l) \rightsquigarrow \delta(r) = \{p : l, \ldots\}$
  The triple states that resource r is related to literal l with property p. Both the property and literal are descriptors, so they must be combined into a composite descriptor, similarly to the example of previous section. We again use an open set because RDFS properties can be multi-valued, and also because resources also have classes as descriptors.

To summarize, the set of descriptions can be defined as follows.

$$D = D_1 \cup D_2 \quad \text{where} \quad D_1 = 2^{\mathcal{C}} \times 2^{\mathcal{P} \times \mathcal{L}} \quad \text{and} \quad D_2 = 2^{\mathcal{P}}$$

D_1 is the set of descriptions of individuals resources, where a description is a pair made of a set of classes, and a set of valued properties. D_2 is the set of descriptions of edges between resources, where a description is a set of properties. The empty description is therefore simply the empty set: $\bot = \emptyset$. From there, we can formally define the description of every resources and pairs of resources.

$$\delta(r) = (\{c \mid (r, \mathtt{rdf:type}, c) \in T\}, \{p\!:\!l \mid (r, p, l) \in T, \ l \in \mathcal{L}\})$$
$$\delta(r, r') = \{p \mid (r, p, r') \in T, \ r' \in R\}$$

It remains to define the similarity operator $\sqcap$ over descriptions. As our descriptions of RDFS resources are based on sets of elementary descriptors, we derive similarity on sets from similarity on elements. We can allow the similarity between two elements to be a set of elements, the *least general generalizations* *(lgg)*. This set-based approach has already been used in Pattern Structures, e.g. for graphs and subgraphs [16]. On classes and properties, we have a partial ordering $\leq$ from which the *lgg* operator can defined as follows:

$$lgg(x, y) := Min_{\leq}\{z \in X \mid x \leq z, y \leq z\}$$

On literals, given that $\mathcal{L}$ is in general infinite, it is more convenient to assume the *lgg* operation to be defined, and to derive the partial ordering from it: $x \leq y \iff lgg(x, y) = \{y\}$. Here are a few examples on how *lgg* could be defined on literals:

- $lgg(10, 20) = \{[10, 20]\}$,
- $lgg([10, 30], [25, 50]) = \{[10, 50]\}$,
- $lgg(\text{"Formal Concept Analysis"}, \text{"Relational Concept Analysis"}) = \{\text{"Concept"}, \text{"Analysis"}\}$.

Of course, this assumes to extend the set of literals $\mathcal{L}$ with all patterns that may be generated by the *lgg* operator, e.g. intervals. On valued properties the *lgg* operator can be obtained by combining the *lgg* operators on properties and literals.

$$lgg(p'\!:\!l', p''\!:\!l'') = \{p\!:\!l \mid p' \in lgg(p', p''), l \in lgg(l', l'')\}$$

Each *lgg* operator can be lifted to the PS similarity operator on sets of elements by collecting all least general generalizations of elements pairwise, and then filtering them to keep only the most specific ones.

$$d' \sqcap d'' = Min_{\leq}\{x \in lgg(x', x'') \mid x' \in d', y' \in d''\}$$

This is enough to define similarity on D_2-descriptions, which are sets of properties. On D_1-descriptions, similarity can be defined element-wise as they are pairs (C, PL) of sets of elementary descriptors: C is a set of classes, and PL is a set of valued properties.

$$(C', PL') \sqcap (C'', PL'') = (C' \sqcap C'', PL' \sqcap PL'')$$

Finally, the similarity between a D_1-description and a D_2-description is simply the empty description $\perp$, although Graph-PS only applies similarity to the descriptions of tuples of objects with the same arity.

5   Conclusion and Perspectives

We have introduced a new extension of Formal Concept Analysis that merges two existing FCA extensions, Pattern Structures (PS) and Graph-FCA. In short, PS-like descriptions are used to describe the nodes and hyperedges of graphs, in place of sets of attributes. The new extension therefore combines the benefits of the two existing extensions: complex descriptions and relationships between objects. A strength of Graph-PS is that it is a proper generalization of PS and Graph-FCA, in the sense that PS and Graph-FCA – as well as FCA – are special cases of Graph-PS. Hence, all previous work about defining custom pattern structures can be reused in Graph-PS, and the compact graphical representations of concept lattices in Graph-FCA can be reused in Graph-PS. We have also shown that Graph-PS can accurately represent existing graph-based models like RDFS graphs.

This paper focuses on the theoretical aspects of Graph-PS, and the most immediate perspectives concern its implementation and its applications. The implementation could be adapted from the existing implementation of Graph-FCA [7], by taking into account the similarity operator $\sqcap$ in the PGP operations $\sqsubseteq_q$ and $\sqcap_q$. The additional cost of using Graph-PS in PS and Graph-FCA settings should be evaluated. A toolbox of components should be built in order to facilitate the design of new sets of descriptions, by capitalizing on previous applications of pattern structures, and by adopting the methodology of logic functors [13]. In the end, we plan to experiment Graph-PS in diverse knowledge graphs and other complex structures like sequences and trees.

References

1. Alam, M., Buzmakov, A., Napoli, A.: Exploratory knowledge discovery over web of data. Discret. Appl. Math. **249**, 2–17 (2018)
2. Ayats, H., Cellier, P., Ferré, S.: Extracting relations in texts with concepts of neighbours. In: Braud, A., Buzmakov, A., Hanika, T., Le Ber, F. (eds.) ICFCA 2021. LNCS (LNAI), vol. 12733, pp. 155–171. Springer, Cham (2021). https://doi.org/10.1007/978-3-030-77867-5_10
3. Baader, F., Calvanese, D., McGuinness, D.L., Nardi, D., Patel-Schneider, P.F. (eds.): The Description Logic Handbook: Theory, Implementation, and Applications. Cambridge University Press, New York (2003)
4. Baixeries, J., Kaytoue, M., Napoli, A.: Characterizing functional dependencies in formal concept analysis with pattern structures. Ann. Math. Artif. Intell. **72**, 129–149 (2014)
5. Belohlavek, R.: Fuzzy closure operators. J. Math. Anal. Appl. **262**, 473–489 (2001)
6. Chaudron, L., Maille, N.: Generalized formal concept analysis. In: Ganter, B., Mineau, G.W. (eds.) ICCS-ConceptStruct 2000. LNCS (LNAI), vol. 1867, pp. 357–370. Springer, Heidelberg (2000). https://doi.org/10.1007/10722280_25
7. Ferré, S., Cellier, P.: Modeling complex structures in Graph-FCA: illustration on natural language syntax. In: Existing Tools and Applications for Formal Concept Analysis (ETAFCA), pp. 1–6 (2022)
8. Ferré, S.: A proposal for extending formal concept analysis to knowledge graphs. In: Baixeries, J., Sacarea, C., Ojeda-Aciego, M. (eds.) ICFCA 2015. LNCS (LNAI),

vol. 9113, pp. 271–286. Springer, Cham (2015). https://doi.org/10.1007/978-3-319-19545-2_17

9. Ferré, S.: Answers partitioning and lazy joins for efficient query relaxation and application to similarity search. In: Gangemi, A., et al. (eds.) ESWC 2018. LNCS, vol. 10843, pp. 209–224. Springer, Cham (2018). https://doi.org/10.1007/978-3-319-93417-4_14

10. Ferré, S., Cellier, P.: Graph-FCA in practice. In: Haemmerlé, O., Stapleton, G., Faron Zucker, C. (eds.) ICCS 2016. LNCS (LNAI), vol. 9717, pp. 107–121. Springer, Cham (2016). https://doi.org/10.1007/978-3-319-40985-6_9

11. Ferré, S., Cellier, P.: Graph-FCA: an extension of formal concept analysis to knowledge graphs. Discret. Appl. Math. **273**, 81–102 (2019)

12. Ferré, S., Ridoux, O.: A logical generalization of formal concept analysis. In: Ganter, B., Mineau, G.W. (eds.) ICCS-ConceptStruct 2000. LNCS (LNAI), vol. 1867, pp. 371–384. Springer, Heidelberg (2000). https://doi.org/10.1007/10722280_26

13. Ferré, S., Ridoux, O.: A framework for developing embeddable customized logics. In: Pettorossi, A. (ed.) LOPSTR 2001. LNCS, vol. 2372, pp. 191–215. Springer, Heidelberg (2002). https://doi.org/10.1007/3-540-45607-4_11

14. Ferré, S.: Application of concepts of neighbours to knowledge graph completion. Data Sci. Methods Infrastruct. Appl. **4**, 1–28 (2021)

15. Ferré, S., Huchard, M., Kaytoue, M., Kuznetsov, S.O., Napoli, A.: Formal concept analysis: from knowledge discovery to knowledge processing. In: Marquis, P., Papini, O., Prade, H. (eds.) A Guided Tour of Artificial Intelligence Research, pp. 411–445. Springer, Cham (2020). https://doi.org/10.1007/978-3-030-06167-8_13

16. Ganter, B., Kuznetsov, S.O.: Pattern structures and their projections. In: Delugach, H.S., Stumme, G. (eds.) ICCS-ConceptStruct 2001. LNCS (LNAI), vol. 2120, pp. 129–142. Springer, Heidelberg (2001). https://doi.org/10.1007/3-540-44583-8_10

17. Ganter, B., Wille, R.: Formal Concept Analysis. Springer, Heidelberg (1999). https://doi.org/10.1007/978-3-642-59830-2

18. Kaytoue, M., Kuznetsov, S.O., Napoli, A.: Revisiting numerical pattern mining with formal concept analysis. In: International Joint Conference on Artificial Intelligence (IJCAI) (2011)

19. Kötters, J.: Concept lattices of a relational structure. In: Pfeiffer, H.D., Ignatov, D.I., Poelmans, J., Gadiraju, N. (eds.) ICCS-ConceptStruct 2013. LNCS (LNAI), vol. 7735, pp. 301–310. Springer, Heidelberg (2013). https://doi.org/10.1007/978-3-642-35786-2_23

20. Kuznetsov, S.O., Samokhin, M.V.: Learning closed sets of labeled graphs for chemical applications. In: Kramer, S., Pfahringer, B. (eds.) ILP 2005. LNCS (LNAI), vol. 3625, pp. 190–208. Springer, Heidelberg (2005). https://doi.org/10.1007/11536314_12

21. Lehmann, F., Wille, R.: A triadic approach to formal concept analysis. In: Ellis, G., Levinson, R., Rich, W., Sowa, J.F. (eds.) ICCS-ConceptStruct 1995. LNCS, vol. 954, pp. 32–43. Springer, Heidelberg (1995). https://doi.org/10.1007/3-540-60161-9_27

22. Poelmans, J., Ignatov, D.I., Kuznetsov, S.O., Dedene, G.: Fuzzy and rough formal concept analysis: a survey. Int. J. General Systems **43**(2), 105–134 (2014)

23. Rouane-Hacene, M., Huchard, M., Napoli, A., Valtchev, P.: Relational concept analysis: mining concept lattices from multi-relational data. Ann. Math. Artif. Intell. **67**(1), 81–108 (2013)

24. Voutsadakis, G.: Polyadic concept analysis. Order **19**, 295–304 (2002)

On the Commutative Diagrams Among Galois Connections Involved in Closure Structures

Manuel Ojeda-Hernández[(✉)] [ID], Inma P. Cabrera [ID], Pablo Cordero [ID],
and Emilio Muñoz-Velasco [ID]

Universidad de Málaga, Andalucía Tech, Málaga, Spain
manuojeda@uma.es

Abstract. In previous works we proved that fuzzy closure structures such as closure systems, fuzzy closure systems, fuzzy closure operators and fuzzy closure relations are formal concepts of five Galois connections (three antitone and two isotone), together with the commutativity of the fuzzy part of the diagram. In this work we study the commutativity of the crisp diagram formed by three of these Galois connections.

1 Introduction

Galois connections appear in several mathematical theories and in many situation related to the theory of relations [17]. In particular, it is well-known that the derivation operators of Formal Concept Analysis form a Galois connection [9]. The extension of the notion of Galois connection to the fuzzy framework was introduced by Bělohlávek [1], which provided a way to study Fuzzy Formal Concept Analysis.

Closure systems are very used in computer science and in both pure and applied mathematics [7]. The fuzzy extension of closure systems has been approached from several distinct perspectives in the literature, to cite a few we mention the following [2,6,8,10,11,14]. In this work, we will use the definition introduced in [14], which extends closure systems as meet-subsemilattices in the framework of complete fuzzy lattices. On the other hand, the so-called closure operators, have also been extended to the fuzzy setting, and most authors use the same definition, i.e., a mapping that is inflationary, isotone and idempotent. Fuzzy closure operators were defined in [2,5] and they appear naturally in different areas of fuzzy logic and its applications.

In this paper, we continue the study started in [13,15]. The topic of [13] was the study of the mappings that relate one closure structure to the other. For example, if $\Phi \in L^A$ is a fuzzy closure system, then the mapping $\widehat{\mathsf{c}}(\Phi) \colon A \to A$ defined as $\widehat{\mathsf{c}}(\Phi)(a) = \bigsqcap(a^\rho \otimes \Phi)$ is a closure operator. This allows us to define $\widehat{\mathsf{c}} \colon L^A \to A^A$ as $\Phi \mapsto \widehat{\mathsf{c}}(\Phi)$ for all $\Phi \in L^A$. Similarly, for a closure operator $\mathsf{c} \colon A \to A$, the mapping $\widetilde{\Psi} \colon A^A \to L^A$ maps c to a fuzzy closure system $\widetilde{\Psi}(\mathsf{c})$, defined by $\widetilde{\Psi}(\mathsf{c})(a) = \rho(\mathsf{c}(a), a)$. One of the main results in [13] proved that the pair $(\widehat{c}, \widetilde{\Psi})$ is a fuzzy Galois connection between (L^A, S) and $(\mathrm{Isot}(A^A), \widetilde{\rho})$. Furthermore, it is proved that any pair of closure structures (Φ, c) is a formal concept of the Galois connection. This problem is also studied for fuzzy closure relations, hence studying two additional Galois connections, one between $(\mathrm{Isot}(A^A), \widetilde{\rho})$ and $(\mathrm{IsotTot}(L^{A \times A}), \widehat{\rho})$ and another one between (L^A, S) and $(\mathrm{IsotTot}(L^{A \times A}), \widehat{\rho})$. In the case of paper [15], the main goal is to insert crisp closure systems in the problem. We proved the existence of crisp Galois connections between the crisp lattice $(2^A, \subseteq)$ and the three sets of the previous paragraph endowed with the 1-cut of their fuzzy relations. Moreover, the existence of Galois connections and the behavior of the sets of formal concepts have been considered.

The main goal of this paper is to study the commutativity of the diagram of the three Galois connections presented in [15]. We prove that, of six possible commutative diagrams, two of them are indeed commutative, another one is commutative in the case where the underlying structure is a Heyting algebra, and the remaining three diagrams are not commutative in general, not even under the Heyting algebra assumption.

The outline of the paper is as follows. First, a section of preliminaries to recall already known results that are useful to understand the paper better. The next section introduces the crisp closure systems to the framework and considers the new Galois connections and their formal concepts. The main results of the paper appear in the following section, where the study of the commutativity of the diagrams is completed. Finally, there is a section of conclusions and further work where the results are discussed and some hints of future research lines are shown.

2 Preliminaries

This section presents the necessary notions and results to properly follow the paper. The general framework throughout the paper is going to be a complete residuated lattice $\mathbb{L} = (L, \wedge, \vee, \times \rightarrow, 0, 1)$. For the properties of residuated lattices we refer the reader to [3, Chapter 2].

Given a $\mathbb{L}$-fuzzy set $X \in L^A$, the 1-cut of X, denoted by X^1, is the crisp set $\{a \in A \mid X(a) = 1\}$. Equivalently, we will consider it as the fuzzy set whose characteristic mapping is $X^1(a) = 1$ if $X(a) = 1$ and 0 otherwise. Given a fuzzy relation μ between A and B, i.e., a crisp mapping $\mu \colon A \times B \rightarrow L$, and $a \in A$, the *afterset* a^μ is the fuzzy set $a^\mu \colon B \rightarrow L$ given by $a^\mu(b) = \mu(a, b)$. A fuzzy relation μ is said to be *total* if, for all $a \in A$, the aftersets a^μ are normal fuzzy sets, i.e., there exists $x \in A$ such that $a^\kappa(x) = 1$.

For ρ being a binary $\mathbb{L}$-relation in A, we say that

- ρ is *reflexive* if $\rho(x, x) = 1$ for all $x \in A$.
- ρ is *symmetric* if $\rho(x, y) = \rho(y, x)$ for all $x, y \in A$.
- ρ is *antisymmetric* if $\rho(x, y) \otimes \rho(y, x) = 1$ implies $x = y$ for all $x, y \in A$.
- ρ is *transitive* if $\rho(x, y) \otimes \rho(y, z) \leq \rho(x, z)$ for all $x, y, z \in A$.

Definition 1. *Given a non-empty set A and a binary $\mathbb{L}$-relation ρ on A, the pair (A, ρ) is said to be a* fuzzy poset *if ρ is a fuzzy order, i.e. if ρ is reflexive, antisymmetric and transitive.*

A typical example of fuzzy poset is (L^A, S), for any set A. If (A, ρ) is a fuzzy poset, we will also use the so-called *full fuzzy powering* ρ_∞, which is the fuzzy relation on L^A defined as follows: for all $X, Y \in L^A$,

$$\rho_\infty(X, Y) = \bigwedge_{x, y \in A} (X(x) \otimes Y(y)) \to \rho(x, y).$$

Since the structure used in this paper is a complete fuzzy lattice, we require the notion of infimum and supremum in the fuzzy setting. These concepts, originally introduced by Bělohlávek in [4], are defined as follows.

Definition 2. *Let (A, ρ) be a fuzzy poset and $X \in L^A$. The down-cone (resp. up-cone) of X is defined as a fuzzy set with the following membership function.*

$$X_\rho(x) = \bigwedge_{a \in A} X(a) \to \rho(x, a) \left(resp.\ X^\rho(x) = \bigwedge_{a \in A} X(a) \to \rho(a, x) \right).$$

Definition 3. *Let (A, ρ) be a fuzzy poset and $X \in L^A$. An element $a \in A$ is said to be* infimum *(resp.* supremum*) of X if the following conditions hold:*

1. *$X_\rho(a) = 1$ (resp. $X^\rho(a) = 1$).*
2. *$X_\rho(x) \leq \rho(x, a)$ (resp. $X^\rho(x) \leq \rho(a, x)$), for all $x \in A$.*

The fuzzy poset (A, ρ) is said to be a complete fuzzy lattice if $\sqcap X$ and $\sqcup X$ exist for all $X \in L^A$.

Hereinafter, suprema and infima in A will be denoted by $\sqcup$ and $\sqcap$, respectively.

Theorem 1. *An element $a \in A$ is infimum (resp. supremum) of $X \in L^A$ if and only if, for all $x \in A$,*

$$\rho(x, a) = X_\rho(x) \qquad (resp.\ \rho(a, x) = X^\rho(x)).$$

Every fuzzy order induces a symmetric relation, called symmetric kernel relation.

Definition 4. *Given a fuzzy poset (A, ρ), the* symmetric kernel relation *is defined as $\approx: A \times A \to L$ where $(x \approx y) = \rho(x, y) \otimes \rho(y, x)$ for all $x, y \in A$.*

The notion of extensionality was introduced in the very beginning of the study of fuzzy sets. It has also been called compatibility (with respect to the similarity relation) in the literature.

Definition 5. *Let (A, ρ) be a fuzzy poset and $X \in L^A$.*

– *X is said to be* extensional *or* compatible *with respect to $\approx$ if it satisfies $X(x) \otimes (x \approx y) \leq X(y)$, for all $x, y \in A$.*

– The *extensional hull* of X, denoted by $X^{\approx}$, is the smallest extensional set that contains X. Its explicit formula is the following:

$$X^{\approx}(x) = \bigvee_{a \in A} (X(a) \otimes (a \approx x)).$$

For a crisp set $X \subseteq A$, the expression of the extensional hull is simplified since we have

$$X^{\approx}(x) = \bigvee_{a \in A} (X(a) \otimes (a \approx x)) = \bigvee_{a \in X} (a \approx x).$$

We focus now on closure structures in the fuzzy framework. The definition of fuzzy closure operator in the fuzzy setting is the one used in [2,3].

Definition 6. *Given a fuzzy poset (A, ρ), a mapping $c \colon A \to A$ is said to be a closure operator on $\mathbb{A}$ if the following conditions hold:*

1. $\rho(a, b) \leq \rho(c(a), c(b))$, *for all $a, b \in A$ (isotonicity)*
2. $\rho(a, c(a)) = 1$, *for all $a \in A$ (inflationarity)*
3. $\rho(c(c(a)), c(a)) = 1$, *for all $a \in A$ (idempotency)*

An element $q \in A$ is said to be closed *for c if $\rho(c(q), q) = 1$.*

The counterpart of closure operators are the so-called closure systems, there are various distinct approaches to this concept in the fuzzy framework, originally introduced on the fuzzy powerset lattice by Bělohlávek [2]. The extension to arbitrary complete fuzzy lattices was introduced in [14] and is the one used here.

Definition 7. *Let (A, ρ) be a complete fuzzy lattice. A crisp set $\mathcal{F} \subseteq A$ is said to be a closure system if $\bigsqcap X \in \mathcal{F}$ for all $X \in L^{\mathcal{F}}$.*

This definition of closure system in the fuzzy framework maintains the one-to-one relation of the crisp case [11].

Theorem 2. *Let (A, ρ) be a complete fuzzy lattice. The following assertions hold:*

1. *If c is a closure operator on (A, ρ), the crisp set $\mathcal{F}_c$ defined as $\{a \in A \mid c(a) = a\}$ is a closure system.*
2. *If $\mathcal{F}$ is a closure system, the mapping $c_{\mathcal{F}} \colon A \to A$ defined as $c_{\mathcal{F}}(a) = \bigsqcap(a^{\rho} \otimes \mathcal{F})$ is a closure operator on (A, ρ).*
3. *If $c \colon A \to A$ is a closure operator on (A, ρ), then $c_{\mathcal{F}_c} = c$.*
4. *If $\mathcal{F}$ is a closure system, then $\mathcal{F} = \mathcal{F}_{c_{\mathcal{F}}}$.*

Notice that the set $\mathcal{F}_c$ can be defined as $\{a \in A \mid \rho(c(a), a) = 1\}$ since, as c is inflationary, if $\rho(c(a), a) = 1$ we would also have $\rho(a, c(a)) = 1$ and then $c(a) = a$ by antisymmetry.

Since closure systems are crisp structures with certain fuzzy properties, it is natural to wonder whether a fuzzy structure can be defined. An affirmative answer was found in [14].

Definition 8. *Let* (A, ρ) *be a complete fuzzy lattice. A fuzzy set* $\Phi \in L^A$ *is said to be a fuzzy closure system if* Φ^1 *is a closure system and* Φ *is the extensional hull of* Φ^1.

This definition of fuzzy closure system maintains the well-known one-to-one relationship between closure systems and closure operators [14].

Theorem 3. *Let* (A, ρ) *be a complete fuzzy lattice. The following assertions hold:*

1. *If* c *is a closure operator on* (A, ρ), *the fuzzy set* Φ_c *defined as* $\Phi_c(a) = \rho(c(a), a)$ *is a fuzzy closure system.*
2. *If* Φ *is a fuzzy closure system, the mapping* $c_\Phi \colon A \to A$ *defined as* $c_\Phi(a) = \bigsqcap(a^\rho \otimes \Phi)$ *is a closure operator on* (A, ρ).
3. *If* $c \colon A \to A$ *is a closure operator on* (A, ρ), *then* $c_{\Phi_c} = c$.
4. *If* Φ *is a fuzzy closure system on* (A, ρ), *then* $\Phi = \Phi_{c_\Phi}$.

Analogously, the same discussion can be done by extending closure operators to fuzzy closure relations, which are isotone, idempotent and inflationary fuzzy relations $\kappa \colon A \times A \to L$. A fuzzy closure relation that is minimal among the extensional fuzzy closure relations is named strong fuzzy closure relations. These were introduced in [12]. Therein, it was proved the one-to-one correspondence between strong fuzzy closure relations and fuzzy closure systems [12, Theorem 23].

Fuzzy Galois connections are a main concept in this paper as well. Let us recall the definition.

Definition 9. ([18]). *Let* (A, ρ_A) *and* (B, ρ_B) *be fuzzy posets,* $f \colon A \to B$ *and* $g \colon B \to A$ *be two mappings.*

- *The pair* (f, g) *is called an isotone fuzzy Galois connection or fuzzy isotone Galois connection between* (A, ρ_A) *and* (B, ρ_B), *denoted by* $(f, g) \colon (A, \rho_A) \rightleftharpoons (B, \rho_B)$, *if*

$$\rho_A(g(b), a) = \rho_B(b, f(a)) \quad \text{for all } a \in A \text{ and } b \in B.$$

- *The pair* (f, g) *is called a fuzzy Galois connection between* (A, ρ_A) *and* (B, ρ_B), *denoted by* $(f, g) \colon (A, \rho_A) \leftrightharpoons (B, \rho_B)$, *if*

$$\rho_A(a, g(b)) = \rho_B(b, f(a)) \quad \text{for all } a \in A \text{ and } b \in B.$$

A *formal concept*, also called *fixed pair* or *fixed point*, of a fuzzy Galois connection (f, g) is a couple $(a, b) \in A \times B$ such that $f(a) = b$ and $g(b) = a$.

In [13], the mappings in Theorems 2 and 3 and the relationships among them were studied. Specifically, it was proved the existence of fuzzy Galois connections between pairs of these mappings and whether the fuzzy closure structures are formal concepts of them. That is,

$$(L^A, S) \xrightarrow{\ (\widehat{c}, \widetilde{\Psi})\ } (\mathrm{Isot}(A^A), \widetilde{\rho})$$

$$(L^A, S) \xrightarrow{\ (\kappa, \widehat{\Psi})\ } (\mathrm{IsotTot}(L^{A \times A}), \widehat{\rho})$$

$$(\mathrm{Isot}(A^A), \widetilde{\rho}) \xrightleftharpoons{\ (-^1, -^\approx)\ } (\mathrm{IsotTot}(L^{A \times A}), \widehat{\rho}),$$

where $\mathrm{Isot}(A^A)$ is the set of (crisp) isotone mappings on A, $\mathrm{IsotTot}(L^{A\times A})$ is the set of isotone and total fuzzy relations on A and the fuzzy relations $\widetilde{\rho}, \hat{\rho}$ are defined as follows,

$$\widetilde{\rho}(f_1, f_2) = \bigwedge_{x\in A} \rho(f_1(x), f_2(x)) \text{ for all } f_1, f_2 \in A^A$$

$$\hat{\rho}(\kappa_1, \kappa_2) = \bigwedge_{a\in A} \rho_\propto(a^{\kappa_1}, a^{\kappa_2}) \text{ for all } \kappa_1, \kappa_2 \in L^{A\times A}.$$

The mappings in the diagram are defined as follows,

$$\hat{c}\colon L^A \to \mathrm{Isot}(A^A) \qquad\qquad \hat{c}(\Phi) = c_\Phi$$

$$\widetilde{\Psi}\colon \mathrm{Isot}(A^A) \to L^A \qquad\qquad \widetilde{\Psi}(f) = \Phi_f$$

$$\kappa\colon L^A \to \mathrm{IsotTot}(L^{A\times A}) \qquad\qquad \kappa(\Phi)(a,b) = \prod(a^\rho \otimes \Phi) \approx b$$

$$\hat{\Psi}\colon \mathrm{IsotTot}(L^{A\times A}) \to L^A \qquad\qquad \hat{\Psi}(\mu)(x) = \rho_\propto(x^\mu, x)$$

Results in [13] involve fuzzy closure systems, closure operators and strong fuzzy closure relations. Specifically, there we prove the following assertions:

1. If Φ is a fuzzy closure system, $(\Phi, \hat{c}(\Phi))$ is a formal concept of the Galois connection $(\hat{c}, \widetilde{\Psi})$.
2. If c is a fuzzy closure operator, $(\widetilde{\Psi}(c), c)$ is a formal concept of the Galois connection $(\hat{c}, \widetilde{\Psi})$.
3. If Φ is a fuzzy closure system, $(\Phi, \kappa(\Phi))$ is a formal concept of the Galois connection $(\kappa, \hat{\Psi})$.
4. If μ is a strong fuzzy closure relation, $(\hat{\Psi}(\mu), \mu)$ is a formal concept of the Galois connection $(\kappa, \hat{\Psi})$.
5. If c is a fuzzy closure operator, $(c^\approx, c)$ is a formal concept of the fuzzy isotone Galois connection $(-^1, -^\approx)$.
6. If μ is a strong fuzzy closure relation, (μ, μ^1) is a formal concept of the fuzzy isotone Galois connection $(-^1, -^\approx)$.

3   Crisp Closure Systems

Notice that closure systems as crisp sets are not considered in the Galois connections mentioned above. Thus, a natural step is to study a similar problem with the partially ordered set $(2^A, \subseteq)$. This addition to the problem is not straightforward since $(2^A, \subseteq)$ is a crisp poset, whereas (L^A, S) and $(\mathrm{Isot}(A^A), \widetilde{\rho})$ are complete fuzzy lattices and $(\mathrm{IsotTot}(L^{A\times A}), \hat{\rho})$ is a fuzzy preposet [13]. Thus, we will consider the 1-cut of the fuzzy relations defined above and study the crisp problem. It is well-known that S^1 is Zadeh's inclusion, hence throughout the paper we will follow the classical notation for subsethood $\subseteq$. We will also consider the 1-cut of $\widetilde{\rho}$, we will denote $\widetilde{\rho}^1(f, g) = 1$ as $f \preceq g$, and $\hat{\rho}^1(\kappa_1, \kappa_2) = 1$ will be denoted by $\kappa_1 \sqsubseteq \kappa_2$, that is,

$$f \preceq g \text{ if and only if } \rho(f(a), g(a)) = 1, \text{ for all } a \in A. \tag{1}$$

$$\kappa_1 \sqsubseteq \kappa_2 \text{ if and only if } x^{\kappa_1}(y) \otimes x^{\kappa_2}(z) \le \rho(y, z) \text{ for all } x, y, z \in A. \quad (2)$$

Observe that the set $(\mathrm{IsotTot}(L^{A \times A}), \sqsubseteq)$ is a preordered set. In addition, $(L^A, \subseteq)$ and $(\mathrm{Isot}(A^A), \preceq)$ are complete lattices by [13, Proposition 13] and [3, Theorem 4.55].

Proposition 1. *The following pairs of mappings form two Galois connections and an isotone Galois connection, respectively,*

$$(L^A, \subseteq) \xleftarrow{\quad (\widetilde{c}, \widetilde{\Psi}) \quad} (\mathrm{Isot}(A^A), \preceq)$$

$$(L^A, \subseteq) \xleftarrow{\quad (\kappa, \widehat{\Psi}) \quad} (\mathrm{IsotTot}(L^{A \times A}), \sqsubseteq)$$

$$(\mathrm{IsotTot}(L^{A \times A}), \sqsubseteq) \xrightleftharpoons{\quad (-^1, -^{\approx}) \quad} (\mathrm{Isot}(A^A), \preceq).$$

The question now is whether it is possible to consider $(2^A, \subseteq)$ in this problem. This is the topic discussed in [15]. Following the spirit in [13], the mappings which form the new Galois connections must be the ones relating closure systems and fuzzy closure operators, i.e., $\mathcal{F}(f) = \{a \in A \mid \rho(f(a), a) = 1\}$ and $\widetilde{c}(X)(a) = \bigsqcap(a^\rho \otimes X)$, for any isotone mapping $f \in \mathrm{Isot}(A^A)$ and any set $X \subseteq A$. Thus, the couple $(\widetilde{c}, \mathcal{F})$ forms a Galois connection.

$$(2^A, \subseteq) \xleftarrow{\quad (\widetilde{c}, \mathcal{F}) \quad} (\mathrm{Isot}(A^A), \preceq)$$

In order to complete the scenario, our next intended goal would be to prove the existence of isotone Galois connection between $(2^A, \subseteq)$ and (L^A, S). Unfortunately, the couple of mappings we have been considering so far, i.e.,

$$\Phi^1 = \{a \in A \mid \Phi(a) = 1\} \text{ and } X^{\approx}(a) = \bigvee_{x \in X}(x \approx a),$$

do not form an isotone Galois connection between these two posets. Even though this pair is not an isotone Galois connection in general, the restriction to extensional fuzzy sets does. Hence, we will work on the set of extensional fuzzy sets of A, denoted by $(\mathrm{Ext}(L^A), S)$. This set maintains many of the good properties of (L^A, S) such as being a complete fuzzy lattice.

As anticipated above, the couple of mappings $(-^1, -^{\approx})$ forms an isotone Galois connection between $(2^A, \subseteq)$ and $(\mathrm{Ext}(L^A), \subseteq)$.

$$(2^A, \subseteq) \xrightleftharpoons{\quad (-^1, -^{\approx}) \quad} (L^A, \subseteq)$$

Restricting to extensional fuzzy sets maintains the Galois connections in Proposition 1, since the images of both $\widetilde{\Psi}$ and $\widehat{\Psi}$ are always extensional sets and any isotone and total relation is extensional.

The formal concepts of the Galois connection introduced above are studied in the following theorem.

Theorem 4. *Let $X \in 2^A$ and $f \in \text{Isot}(A^A)$. The following statements are equivalent:*

1. *The couple (X, f) is a formal concept of the Galois connection $(\widetilde{c}, \mathcal{F})$.*
2. *The crisp set X is a closure system and $\widetilde{c}(X) = c_X = f$.*
3. *The mapping f is a closure operator and $\mathcal{F}(f) = \mathcal{F}_f = X$.*

The last result shows that every formal concept of the Galois connection is a pair of fuzzy closure structures. This is interesting because this result does not hold in the general fuzzy setting. Recall that in the general fuzzy setting fuzzy closure structures were formal concepts of the Galois connections but there existed formal concepts that were not formed by fuzzy closure structures. Examples of such cases can be found in [13].

As done in the previous case, a study of the formal concepts of the isotone Galois connection $(-^1, -^\approx)$ is necessary. This is done in the result below.

Theorem 5. *Let A be a complete fuzzy lattice, then*

1. *Let $\mathcal{F} \subseteq A$ be a closure system, then $(\mathcal{F}^\approx, \mathcal{F})$ is a formal concept of the isotone Galois connection.*
2. *Let $\Phi \in L^A$ be a fuzzy closure system, then (Φ, Φ^1) is a formal concept of the isotone Galois connection.*

As expected, fuzzy closure structures are formal concepts of the isotone Galois connection. However, there are formal concepts which are not fuzzy closure structures.

4 A Study on Commutativity of Diagrams

The purpose of this section is to examine under which conditions the diagram of all the fuzzy Galois connections in the last section is commutative. The goal is to form a diagram like the following.

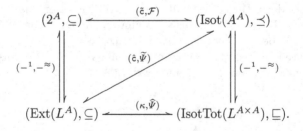

Unfortunately, the diagram is not commutative in general. In [13] an example showing $\kappa \circ \widetilde{\Psi} \neq -^\approx$ and $\widehat{c} \circ \widehat{\Psi} \neq -^1$ was presented.

Some partial results in [13] showed that the following diagrams commute.

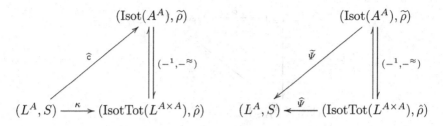

We consider now restrictions to ensure the commutativity of, at least, the upper part of the diagram. Let $\Phi \in L^A$, then $\hat{c}(\Phi)$ is not a closure operator in general. However, $\tilde{c}(\Phi^1)$ will be a closure operator due to Theorem 4. Thus, as expected, the diagram is not commutative in general. In the rest of this section we will give partial results on the relationships among the different compositions.

Lemma 1. *Let $X \in 2^A$ and $\Phi \in \mathrm{Ext}(L^A)$. Then,*

$$\hat{c}(X^{\approx}) \preceq \tilde{c}(X) \qquad and \qquad \hat{c}(\Phi) \preceq \tilde{c}(\Phi^1)$$

Proof. For all $X \subseteq A$ and $a, x \in A$, we have that

$$
\begin{aligned}
(a^\rho \otimes X^{\approx})_\rho (x) &= \bigwedge_{y \in A} ((a^\rho \otimes X^{\approx})(y) \to \rho(x, y)) \\
&= \bigwedge_{y \in A} ((\rho(a, y) \otimes X^{\approx}(y)) \to \rho(x, y)) \\
&= \bigwedge_{y \in A} \left(\left(\rho(a, y) \otimes \bigvee_{z \in X} (y \approx z) \right) \to \rho(x, y) \right) \\
&\overset{(i)}{=} \bigwedge_{z \in X} \bigwedge_{y \in A} ((\rho(a, y) \otimes (y \approx z)) \to \rho(x, y)) \\
&\overset{(ii)}{\leq} \bigwedge_{z \in X} (\rho(a, z) \to \rho(x, z)) = (a^\rho \otimes X)_\rho (x)
\end{aligned}
$$

where, in (i), we have used [3, (2.50) and (2.52)], and, in (ii), we have considered $y = z$.

Since infima is defined in terms of lower-cones we have that

$$\rho(\hat{c}(X^{\approx})(a), \tilde{c}(X)(a)) = \rho(\textstyle\prod(a^\rho \otimes X^{\approx}), \textstyle\prod(a^\rho \otimes X)) = 1.$$

On the other hand, for all $\Phi \in \mathrm{Ext}(L^A)$ and all $a \in A$, we have that $\Phi^1 \subseteq \Phi$ and, therefore,

$$\rho(\hat{c}(\Phi)(a), \tilde{c}(\Phi^1)(a)) = \rho(\textstyle\prod(a^\rho \otimes \Phi), \textstyle\prod(a^\rho \otimes \Phi^1)) = 1$$

Proposition 2. *The following diagram commutes*

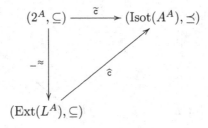

Proof. By Lemma 1, we have that $\hat{c}(X^{\approx}) \preceq \tilde{c}(X)$. Let us prove the converse inequality. Let $X \subseteq A$ and $a, x \in A$. Then,

$$(a^{\rho} \otimes X^{\approx})_{\rho}(x) = \bigwedge_{y \in A} ((a^{\rho} \otimes X^{\approx})(y) \to \rho(x, y))$$

$$= \bigwedge_{y \in A} ((\rho(a, y) \otimes X^{\approx}(y)) \to \rho(x, y))$$

$$= \bigwedge_{y \in A} \left(\left(\rho(a, y) \otimes \bigvee_{z \in X} (X(z) \otimes (y \approx z)) \right) \to \rho(x, y) \right)$$

$$\overset{(i)}{=} \bigwedge_{y, z \in A} ((\rho(a, y) \otimes X(z) \otimes (y \approx z)) \to \rho(x, y))$$

$$\overset{(ii)}{=} \bigwedge_{y, z \in A} ((\rho(a, y) \otimes \rho(y, z) \otimes X(z)) \to (\rho(z, y) \to \rho(x, y)))$$

$$\overset{(iii)}{\geq} \bigwedge_{y, z \in A} ((\rho(a, y) \otimes \rho(y, z) \otimes X(z)) \to \rho(x, z))$$

$$\overset{(iv)}{\geq} \bigwedge_{z \in A} ((\rho(a, z) \otimes X(z)) \to \rho(x, z)) = (a^{\rho} \otimes X)_{\rho}(x),$$

where (i) holds due to [3, (2.50), (2.52)], (ii) holds by [3, (2.33)], (iii) holds by transitivity of ρ and the adjunction and finally (iv) holds due to transitivity and [3, (2.44)].

The following example shows that the inequality $\hat{c}(\Phi) \preceq \tilde{c}(\Phi^1)$ could be strict. Therefore, the corresponding diagram does not commute in general.

Example 1. Let $\mathbb{L} = (\{0, 0.5, 1\}, \wedge, \vee, \otimes, \to, 0, 1)$ be a Łukasiewicz residuated lattice with three values, and (A, ρ) be the fuzzy lattice with $A = \{\bot, a, b, c, d, e, \top\}$

and the fuzzy relation $\rho\colon A \times A \to L$ described by the following table:

ρ	$\bot$	a	b	c	d	e	$\top$
$\bot$	1	1	1	1	1	1	1
a	0.5	1	0.5	1	1	1	1
b	0.5	0.5	1	1	1	1	1
c	0.5	0.5	0.5	1	1	1	1
d	0	0.5	0	0.5	1	0.5	1
e	0	0	0.5	0.5	0.5	1	1
$\top$	0	0	0	0.5	0.5	0.5	1

Consider $\Phi = \{a/0.5, b/0.5\}$. It is clear that $\Phi^1 = \varnothing$ and $\tilde{c}(\Phi^1)$ is the constant mapping which maps every element to $\top$ whereas $\hat{c}(\Phi)(a) = c \neq \top$. This is proved as follows.

$$(a^\rho \otimes \Phi)_\rho(c) = \bigwedge_{x \in A} ((a^\rho \otimes \Phi)(x) \to \rho(c, x))$$
$$= ((a^\rho \otimes \Phi)(a) \to \rho(c, a)) \wedge ((a^\rho \otimes \Phi)(b) \to \rho(c, b)) = 1.$$

On the other hand,

$$(a^\rho \otimes \Phi)_\rho(\top) = \bigwedge_{y \in A} ((a^\rho \otimes \Phi)(y) \to \rho(\top, y))$$
$$= ((a^\rho \otimes \Phi)(a) \to \rho(\top, a)) \wedge ((a^\rho \otimes \Phi)(b) \to \rho(\top, b))$$
$$= (0.5 \to 0) \wedge (0.5 \to 0) = 0.5.$$
$$(a^\rho \otimes \Phi)_\rho(d) = \bigwedge_{y \in A} ((a^\rho \otimes \Phi)(y) \to \rho(d, y)) = (0.5 \to 0.5) \wedge (0.5 \to 0) = 0.5.$$
$$(a^\rho \otimes \Phi)_\rho(e) = \bigwedge_{y \in A} ((a^\rho \otimes \Phi)(y) \to \rho(e, y)) = (0.5 \to 0) \wedge (0.5 \to 0.5) = 0.5.$$

These values of the lower bound, by Theorem 1, imply that $\hat{c}(\Phi)(a) = \bigsqcap(a^\rho \otimes \Phi) = c$, thus $\hat{c}(\Phi) \neq \tilde{c}(\Phi^1)$.

Now we focus on the commutativity of another parts of the diagram under study, that is, we examine whether $\mathcal{F}(f) = (\tilde{\Psi}(f))^1$ or $(\mathcal{F}(f))^\approx = \tilde{\Psi}(f)$ for every isotone mapping $f\colon A \to A$. The answer of the first one is affirmative.

Proposition 3. *The following diagram commutes,*

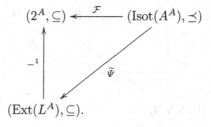

Proof. Let $f\colon A \to A$ be an isotone mapping. Then we have,

$$\tilde{\Psi}(f)(x) = 1 \text{ if and only if } \rho(f(x), x) = 1$$
$$\text{if and only if } f(x) \leq x \text{ if and only if } x \in \mathcal{F}(f).$$

Therefore, $(\tilde{\Psi}(f))^1 = \mathcal{F}(f)$.

Despite the straightforwardness of the last result, the second equality is not that easy to prove. In fact, in the general case we only have one of the inequalities.

Proposition 4. *Let $f\colon A \to A$ be an isotone mapping. Then, $(\mathcal{F}(f))^{\approx} \subseteq \tilde{\Psi}(f)$.*

Proof. Let $f\colon A \to A$ be an isotone mapping. Then we have,

$$\mathcal{F}(f) = \{x \in A \mid f(x) \leq x\}$$
$$(\mathcal{F}(f))^{\approx} = \bigvee_{a \in \mathcal{F}(f)} (x \approx a)$$
$$\tilde{\Psi}(f)(x) = \rho(f(x), x).$$

Let $x \in A, a \in \mathcal{F}(f)$, then we get,

$$
\begin{aligned}
(x \approx a) = \rho(x, a) \otimes \rho(a, x) & \\
\leq \rho(f(x), f(a)) \otimes \rho(a, x) & \qquad \text{by isotonicity of } f \\
= \rho(f(x), f(a)) \otimes \rho(f(a), a) \otimes \rho(a, x) & \qquad \text{by } a \in \mathcal{F}(f) \\
\leq \rho(f(x), x) & \qquad \text{by transitivity.}
\end{aligned}
$$

This argument is valid for all $a \in \mathcal{F}(f)$, therefore $\bigvee_{a \in \mathcal{F}(f)}(x \approx a) \leq \rho(f(x), x)$.

The following example shows that the inclusion $(\mathcal{F}(f))^{\approx} \subseteq \tilde{\Psi}(f)$ could be strict, and, therefore, the corresponding diagram does not commute.

Example 2. Consider the residuated lattice $\mathbb{L}$ and the fuzzy lattice (A, ρ) given in Example 1. Let $f \in \text{Isot}(A^A)$ with $f(\bot) = f(a) = a$, $f(b) = c$, $f(c) = f(d) = d$ and $f(e) = f(\top) = \top$.

On the one hand, we have that $\tilde{\Psi}(f) = \{^{\top}/0.5, {}^{a}/1, {}^{b}/0.5, {}^{c}/0.5, {}^{d}/1, {}^{e}/0.5, {}^{\top}/1\}$.
On the other hand, $\mathcal{F}(f) = \{a, d, \top\}$ and

$$(\mathcal{F}(f))^{\approx} = \{^{\top}/0.5, {}^{a}/1, {}^{b}/0, {}^{c}/0.5, {}^{d}/1, {}^{e}/0.5, {}^{\top}/1\} \neq \tilde{\Psi}(f)$$

Notice that this situation can be improved if the structure of the membership degrees is a Heyting algebra, as the following proposition assures.

Proposition 5. *Let (A, ρ) be a fuzzy lattice over a Heyting algebra $\mathbb{L}$. The following diagram commutes,*

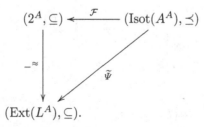

Proof. For all $f \in \text{Isot}(A^A)$, since $\mathbb{L}$ is a Heyting algebra, by [16, Theorem 4], we have that $\widetilde{\Psi}(f)$ is a fuzzy closure system. Then, by definition of fuzzy closure system (Definition 8), we have that $((\widetilde{\Psi}(f))^1)^{\approx} = \widetilde{\Psi}(f)$.

On the other hand, by Proposition 3, we have that $(\widetilde{\Psi}(f))^1 = \mathcal{F}(f)$. Therefore, $(\mathcal{F}(f))^{\approx} = \widetilde{\Psi}(f)$.

Going back to earlier results, Example 1 shows that the equality $\hat{c}(\Phi) = \widetilde{c}(\Phi^1)$ does not hold in general. As a matter of fact, as the following example shows, this equality does not even hold in Heyting algebras.

Example 3. Let us consider $A = \{a, b\}$ and the relation ρ given in the following table

ρ	a	b
a	1	1
b	0.6	1

The pair (A, ρ) is a fuzzy lattice over the Heyting algebra $[0, 1]$. For the set $\Phi = \{a/0.8, b/1\}$ we have that $\hat{c}(\Phi)$ is the identity mapping. However, $\widetilde{c}(\Phi^1)$ is the constant mapping $\widetilde{c}(\Phi^1)(x) = b$ for all $x \in A$. Therefore, even under Heyting algebras, $\hat{c} \neq (\widetilde{c} \circ -^1)$.

Proposition 6. *For all $X \subseteq A$ and $\phi \in L^A$, we have that $X^{\approx} \subseteq \widetilde{\Psi}(\widetilde{c}(X))$ and $\Phi^1 \subseteq \mathcal{F}(\hat{c}(\Phi))$.*

Proof. First, given $X \subseteq A$, we have that $\widetilde{c}(X)(a) = \prod(a^{\rho} \otimes X)$ for all $a \in A$ and, by Theorem 1, it means that $\rho(z, \widetilde{c}(X)(a)) = \bigwedge_{x \in X}(\rho(a, x) \to \rho(z, x))$. Thus, by considering $z = \widetilde{c}(X)(a)$, we have that $\rho(a, x) \leq \rho(\widetilde{c}(X)(a), x)$ for all $x \in X$. Therefore, for all $a \in A$ and $x \in X$, we have that

$$(a \approx x) = \rho(a, x) \otimes \rho(x, a) \leq \rho(\widetilde{c}(X)(a), x) \otimes \rho(x, a) \leq \rho(\widetilde{c}(X)(a), a)$$

Since it is true for all $x \in X$, it means that

$$X^{\approx}(a) = \bigvee_{x \in X} (a \approx x) \leq \rho(\widetilde{c}(X)(a), a) = \widetilde{\Psi}(\widetilde{c}(X))(a)$$

or, equivalently, $X^{\approx} \subseteq \widetilde{\Psi}(\widetilde{c}(X))$.

Consider now $\Phi \in L^A$. For all $a \in A$, since $\hat{c}(\Phi)(a) = \prod(a^{\rho} \otimes \Phi)$, by Theorem 1, it means that $\rho(z, \widetilde{c}(\Phi)(a)) = \bigwedge_{x \in A}((\rho(a, x) \otimes \Phi(x)) \to \rho(z, x))$. Thus, by considering $z = \widetilde{c}(\Phi)(a)$, we have that, for all $x \in A$,

$$1 = \rho(\widetilde{c}(\Phi)(a), \widetilde{c}(\Phi)(a)) \leq (\rho(a, x) \otimes \Phi(x)) \to \rho(\widetilde{c}(\Phi)(a), x)$$

Now, considering $x = a$, we have that $\Phi(a) \leq \rho(\widetilde{c}(\Phi)(a), a)$. Finally, since $\mathcal{F}(\hat{c}(\Phi)) = \{a \in A \mid \rho(\widetilde{c}(\Phi)(a), a) = 1\}$, we conclude that $\Phi^1 \subseteq \mathcal{F}(\hat{c}(\Phi))$.

Finally, the following example shows that the equalities $\widetilde{\Psi}(\widetilde{c}(X)) = X^{\approx}$ and $\mathcal{F}(\hat{c}(\Phi)) = \Phi^1$ do not hold in general, even when the underlying structure of membership degrees is a Heyting algebra.

Example 4. Let (A, ρ) be the fuzzy lattice introduced in the previous example. For $X = \{a\}$ we have that $\tilde{c}(X)$ is the identity mapping A and $\tilde{\Psi}(\tilde{c}(X)) = A$. However,

$$X^{\approx} = \{a/1, b/0.6\} \neq A = \tilde{\Psi}(\tilde{c}(X)).$$

On the other hand, for $\Phi = \{a/0.8, b/1\}$ we have that $\hat{c}(\Phi)$ is the identity mapping and $\mathcal{F}(\hat{c}(\Phi)) = A$, but $\Phi^1 = \{b\}$.

5   Conclusions and Further Work

This paper continues the line of work which initiated in [13], where the mappings that relate fuzzy closure structures are studied from the point of view of fuzzy Galois connections, together with the commutativity of the corresponding diagrams. In [15], the crisp powerset lattice was introduced in that same framework. In this paper, we have completed following commutative diagrams.

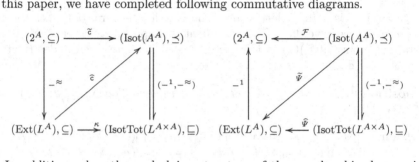

In addition, when the underlying structure of the membership degrees is a Heyting algebra, the second diagram can be improved, i.e. the following diagram commute:

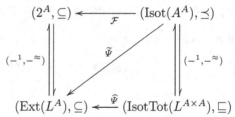

As a prospect of future work, we are considering the extension of this work to more general structures, such as fuzzy posets and trellises.

Acknowledgment. This research is partially supported by the State Agency of Research (AEI), the Spanish Ministry of Science, Innovation, and Universities (MCIU), the European Social Fund (FEDER) and the Universidad de Málaga (UMA) through the FPU19/01467 (MCIU) internship and the research project with reference PID2021-127870OB-I00 (MCIU/AEI/FEDER, UE).

References

1. Bělohlávek, R.: Fuzzy Galois connections. Math. Log. Q. **45**, 497–504 (1999)
2. Bělohlávek, R.: Fuzzy closure operators. J. Math. Anal. Appl. **262**, 473–489 (2001)
3. Bělohlávek, R.: Fuzzy Relational Systems. Springer, New York (2002). https://doi.org/10.1007/978-1-4615-0633-1
4. Bělohlávek, R.: Concept lattices and order in fuzzy logic. Ann. Pure Appl. Logic **128**(1), 277–298 (2004)
5. Bělohlávek, R., De Baets, B., Outrata, J., Vychodil, V.: Computing the lattice of all fixpoints of a fuzzy closure operator. IEEE Trans. Fuzzy Syst. **18**(3), 546–557 (2010)
6. Biacino, L., Gerla, G.: Closure systems and L-subalgebras. Inf. Sci. **33**(3), 181–195 (1984)
7. Caspard, N., Monjardet, B.: The lattices of closure systems, closure operators, and implicational systems on a finite set: a survey. Discret. Appl. Math. **127**(2), 241–269 (2003)
8. Fang, J., Yue, Y.: L-fuzzy closure systems. Fuzzy Sets Syst. **161**(9), 1242–1252 (2010)
9. Ganter, B., Wille, R.: Formal Concept Analysis: Mathematical Foundation. Springer, Heidelberg (1999). https://doi.org/10.1007/978-3-642-59830-2
10. Liu, Y.-H., Lu, L.-X.: L-closure operators, L-closure systems and L-closure L-systems on complete L-ordered sets. In: 2010 Seventh International Conference on Fuzzy Systems and Knowledge Discovery, vol. 1, pp. 216–218 (2010)
11. Ojeda-Hernández, M., Cabrera, I.P., Cordero, P., Muñoz-Velasco, E.: On (fuzzy) closure systems in complete fuzzy lattices. In: 2021 IEEE International Conference on Fuzzy Systems (FUZZ-IEEE), pp. 1–6 (2021)
12. Ojeda-Hernández, M., Cabrera, I.P., Cordero, P., Muñoz-Velasco, E.: Fuzzy closure relations. Fuzzy Sets Syst. **450**, 118–132 (2022)
13. Ojeda-Hernández, M., Cabrera, I.P., Cordero, P., Muñoz-Velasco, E.: Fuzzy closure structures as formal concepts. Fuzzy Sets Syst. (2022). https://doi.org/10.1016/j.fss.2022.12.014
14. Ojeda-Hernández, M., Cabrera, I.P., Cordero, P., Muñoz-Velasco, E.: Fuzzy closure systems: motivation, definition and properties. Int. J. Approx. Reason. **148**, 151–161 (2022)
15. Ojeda-Hernández, M., Cabrera, I.P., Cordero, P., Muñoz-Velasco, E.: Fuzzy closure structures as formal concepts II. Fuzzy Sets and Systems (2023). (Submitted)
16. Ojeda-Hernández, M., Cabrera, I.P., Cordero, P., Muñoz-Velasco, E.: Fuzzy closure systems over Heyting algebras as fixed points of a fuzzy Galois connection. In: Proceedings of the Sixteenth International Conference on Concept Lattices and Their Applications (CLA 2022), vol. 3308, CEUR Workshop Proceedings, pp. 9–18. CEUR-WS.org (2022)
17. Ore, O.: Galois connexions. Trans. Am. Math. Soc. **55**(3), 493–513 (1944)
18. Yao, W., Lu, L.X.: Fuzzy Galois connections on fuzzy posets. Math. Log. Q. **55**, 105–112 (2009)

Scaling Dimension

Bernhard Ganter[1], Tom Hanika[2,3], and Johannes Hirth[2,3(✉)]

[1] TU Dresden, Dresden, Germany
bernhard.ganter@tu-dresden.de
[2] Knowledge and Data Engineering Group, University of Kassel, Kassel, Germany
{tom.hanika,hirth}@cs.uni-kassel.de
[3] Interdisciplinary Research Center for Information System Design,
University of Kassel, Kassel, Germany

Abstract. Conceptual Scaling is a useful standard tool in Formal Concept Analysis and beyond. Its mathematical theory, as elaborated in the last chapter of the FCA monograph, still has room for improvement. As it stands, even some of the basic definitions are in flux. Our contribution was triggered by the study of concept lattices for tree classifiers and the scaling methods used there. We extend some basic notions, give precise mathematical definitions for them and introduce the concept of scaling dimension. In addition to a detailed discussion of its properties, including an example, we show theoretical bounds related to the order dimension of concept lattices. We also study special subclasses, such as the ordinal and the interordinal scaling dimensions, and show for them first results and examples.

Keywords: Formal Concept Analysis · Data Scaling ·
Conceptual Scaling · Ferrers Dimension · Measurement ·
Preprocessing · Feature Compression · Closed Pattern Mining

1 Introduction

When heterogeneous data needs to be analyzed conceptually, e.g., for (closed) pattern mining [11], ontology learning [1] or machine learning [5,10], conceptual scaling [2] is a tool of choice. The task of this method is to translate given data into the standard form, that of a formal context [3]. There are many ways to do this, but not all methods are meaningful in every situation. Often the data has an implicit structure that should guide scaling. For example, it is natural to analyze ordinal data also ordinally. We propose the notion of a *pre-scaling* to reveal such implicit assumptions and to make them usable for a scaling.

Another important aspect is the complexity of the conceptual structure created by scaling. Several authors have suggested only consider important attribute combinations [5,7,12]. We formalize this, speaking of conceptual *views* of the data. It turns out that such views have a natural characterization in terms of *scale measures*, i.e., continuous maps with respect to closure systems that are represented by means of formal contexts. This in turn opens the door to basic

D. Dürrschnabel and D. López Rodríguez (Eds.): ICFCA 2023, LNAI 13934, pp. 64–77, 2023.
https://doi.org/10.1007/978-3-031-35949-1_5

theory questions. We address one of them here for the first time: the question of the *scaling dimension*, i.e., the size of the simplest data set that has the present conceptual structure as its derivative. We study this for the case of ordinal and in particular for that of interordinal scaling, proving characterizations and showing small examples. In addition to our theoretical findings, we demonstrate the applicability of the scaling dimension based on the drive concepts data set and provide all used functions in the `conexp-clj` [6] tool.

2   Formal Concepts Derived from Data Tables

At first glance, it seems very limiting that Formal Concept Analysis focuses on a single basic data type, that of a binary relation between two sets (a *formal context*). Data comes in many different formats, so why restrict to one type only? But this limitation is intentional. It allows a cleaner separation of objective formal data analysis and subjective interpretation, and it allows a unified, clear structure of the mathematical theory.

FCA handles the many different data formats in a two-step process. First, the data is transformed into the standard form – that is, into a formal context – and in the second step that context is analyzed conceptually. The first step, called *conceptual scaling*,[1] is understood as an act of interpretation and depends on subjective decisions of the analyst, who must reveal how the data at hand is meant. It is therefore neither unambiguous nor automatic, and usually does not map the data in its full complexity. However, it is quite possible that several such *conceptual views* together completely reflect the data.

In machine learning the classification data that is used for classifiers are usually lists of n-tuples, as in relational databases or in so-called data tables. In FCA terminology, one speaks of a *many-valued context*. In such a many-valued context, the rows have different names (thereby forming a key), and so do the columns, whose names are called *(many-valued) attributes*. The entries in the table are the values.

Formally, a **many-valued context** $\mathbb{D} := (G, M, W, I)$ consists of a set G of **objects**, a set M of **many-valued attributes**, a set W of **attribute values**, and a ternary relation $I \subseteq G \times M \times W$ satisfying

$$(g, m, v) \in I \text{ and } (g, m, w) \in I \implies v = w.$$

This condition ensures that there is at most one value for each object-attribute pair. The absence of values is allowed. If no values are missing, then one speaks of a **complete** many-valued context. The value for the object-attribute pair

---

[1] The word "scaling" refers to measurement theory, not whether algorithms can be applied to large data sets.

(g, m), if present, sometimes is denoted by $m(g)$, and $m(g) = \bot$ indicates that the value is missing.

It was mentioned above that in order to derive a conceptual structure from a many-valued context, Formal Concept Analysis requires the interpretative step of *conceptual scaling*, which determines the concept-forming attributes which are derived from the many-valued context. Such a formalism is required for this is illustrated by the following example. Suppose that one of the many-valued attributes is "size", with values "very small", "small", "large", "very large". A simple approach would be to use the values as attribute names, introducing attributes "size = very small", "size = small", etc., with the obvious interpretation: an object g has the attribute "size = small" if $m(g) = $ small when $m = $ size. In this *nominal* interpretation the attribute extent for the attribute "size = small" contains those objects which are "small", but excludes those which are "very small", which perhaps is not intended. To repair this, one could use the implicit order

very small < small < large < very large

of the values and derived attributes such as size $\leq$ large, etc. But this *interordinal* interpretation can equally lead to undesired attribute extents, since the attribute size $\leq$ large applies to all object which are "very small", "small", or "large", but excludes those for which no value was noted because their size was unremarkable, neither small nor large. A *biordinal* interpretation can take this into account [3].

2.1 Pre-scalings

Some data tables come with slightly richer information, for which we introduce an additional definition. A **pre-scaling** of a many-valued context $\mathbb{D} := (G, M, W, I)$ is a family $(W(m) \mid m \in M)$ of sets $W(m) \subseteq W$ such that $W = \bigcup_{m \in M} W(m)$ and

$$(g, m, w) \in I \implies w \in W(m)$$

for all $g \in G, m \in M$. We call $W(m)$ the **value domain** of the many-valued attribute m. A tuple $(v_m \mid m \in M)$ **matches** a pre-scaling iff $v_m \in W(m) \cup \{\bot\}$ holds for all $m \in M$. $(G, M, W(m)_{m \in M}, J)$ may be called a **stratified** many-valued context.

It is also allowed that the value domains additionally carry a structure, e.g., are ordered. This also falls under the definition of "pre-scaling". We remain a little vague here, because its seems premature to give a sharp definition. Prediger [13] suggests the notion of a *relational* many-valued context. This may be formalized as a tuple

$$(G, M, (W(m), \mathcal{R}_m)_{m \in M}, I),$$

where $(G, M, W(m)_{m \in M}, I)$ is a stratified many valued-context as defined above, where on each value domain $W(m)$ a family $\mathcal{R}_m$ of relations is given. Prediger and Stumme [12] then discuss deriving one-valued attributes using expressions in a suitable logical language, such as one of the OWL-variants. They call this *logical scaling*.

2.2   Interordinal Plain Scaling

A **scale** for an attribute m of a many-valued context $(G, M, W(m)_{m \in M}, I)$ is a formal context $\mathbb{S}_m := (G_m, M_m, I_m)$ with $W(m) \subseteq G_m$. The objects of a scale are the **scale values**, the attributes are called **scale attributes**.

By specifying a scale a data analyst determines how the attribute values are used conceptually.

For **plain scaling** a formal context (G, N, J) is **derived** from the many-valued context $(G, M, W(m)_{m \in M}, I)$ and the scale contexts $\mathbb{S}_m$, $m \in M$, as follows:

- The object set is G, the same as for the many-valued context,
- the attribute set is the disjoint union of the scale attribute sets, formally

$$N := \bigcup_{m \in M} \{m\} \times M_m,$$

- and the incidence is given by

$$g \, J \, (m, n) :\Longleftrightarrow (g, m, v) \in I \text{ and } v \, I_m \, n.$$

The above definition may look technical, but what is described is rather simple: Every column of the data table is replaced by several columns, one for each scale attribute of $\mathbb{S}_m$, and if the cell for object g and many-valued attribute m contains the value v, then that is replaced by the corresponding "row" of the scale. Choosing the scales is already an act of interpretation, deriving the formal context when scales are given is deterministic.

Pre-scaling, as mentioned above, may suggest the scales to use. An ordered pre-scaling naturally leads to an *interordinal* interpretation of data, using only interordinal scales. We repeat the standard definition of interordinal scaling:

Definition 1 (Interordinal Scaling of $\mathbb{D}$).   *When $\mathbb{D} := (G, M, W, I)$ is a many-valued context with linearly ordered value sets $(W(m), \leq_m)$, then the formal context $\mathbb{I}(\mathbb{D})$ derived from interordinal scaling has G as its object set and attributes of the form*

$$(m, \leq_m, v) \text{ or } (m, \geq_m, v),$$

where v is a value of the many valued attribute m. The incidence is the obvious one, an object g has e.g., the attribute $(m, \leq_m, v)$ iff the value of m for the object g is $\leq_m v$. Instead of $(m, \leq_m, v)$ or $(m, \geq_m, v)$ one writes $m: \leq v$ and $m: \geq v$, respectively. Formally $\mathbb{I}(\mathbb{D}) := (G, N, J)$, where

$$N := \{m: \leq v \mid m \in M, v \in W(m)\} \cup \{m: \geq v \mid m \in M, v \in W(m)\}$$

and

$$(g, m: \leq v) \in J :\Longleftrightarrow m(g) \leq v, \qquad (g, m: \geq v) \in J :\Longleftrightarrow m(g) \geq v.$$

For simplicity, attributes which apply to all or to no objects are usually omitted.

Remark: The formal context derived from a many-valued context $\mathbb{D}$ with linearly ordered value sets via *ordinal plain scaling* is denoted $\mathbb{O}(\mathbb{D})$.

2.3   Scale Measures and Views

By definition, the number of attributes of a derived context is the sum of the numbers of attributes of the scales used, and thus tends to be large. It is therefore common to use selected subsets of these derived attributes, or attribute combinations. This leads to the notion of a *view*:

Definition 2. *A **view** of a formal context (G, M, I) is a formal context (G, N, J), where for each $n \in N$ there is a set $A_n \subseteq M$ such that*

$$g \, J \, n :\Longleftrightarrow A_n \subseteq g^I.$$

*A **contextual view** of a many-valued context $\mathbb{K}$ is a view of a derived context of $\mathbb{K}$; the concept lattice of such a contextual view is a **conceptual view** of $\mathbb{K}$.*

In order to compare contexts derived by conceptual scaling, the notion of a *scale measure* is introduced.

Definition 3. *Let $\mathbb{K} := (G, M, I)$ and $\mathbb{S} := (G_\mathbb{S}, M_\mathbb{S}, I_\mathbb{S})$ be formal contexts. A mapping*

$$\sigma : G \to G_\mathbb{S}$$

*is called an $\mathbb{S}$-**measure** of $\mathbb{K}$ if the preimage $\sigma^{-1}(E)$ of every extent E of $\mathbb{S}$ is an extent of $\mathbb{K}$. An $\mathbb{S}$-measure is **full** if every extent of $\mathbb{K}$ is the preimage of an extent of $\mathbb{S}$.*

Proposition 1. *A formal context $\mathbb{K}_1 := (G, N, J)$ is a view of $\mathbb{K} := (G, M, I)$ if and only if the identity map is a $\mathbb{K}_1$-measure of $\mathbb{K}$.*

Proof. When (G, N, J) is a view of (G, M, I), then every extent E of (G, N, J) is of the form $E = B^J$ for some $B \subseteq N$. Then E also is an extent of (G, M, I), since $E = (\bigcup_{n \in B} A_n)^I$. Conversely, if the identity map is a (G, N, J)-measure of (G, M, I), then for each $n \in N$ the preimage of its attribute extent n^J (which, of course, is equal to n^J) must be an extent of (G, M, I) and therefore be of the form A_n^I for some set $A_n \subseteq M$.   □

As shown by Proposition 1 there is a close tie between contextual views and canonical representation of scale-measures as proposed in Proposition 10 of Hanika and Hirth [7]. Said representations provide for a scale-measure σ of $\mathbb{K}$ into $\mathbb{S}$ an equivalent scale-measure based on the identity map of $\mathbb{K}$ into a context of the form $(G, \mathcal{A}, \in)$ where $\mathcal{A} \subseteq \text{Ext}(\mathbb{K})$, i.e., the set of extents. With the now introduced notions we understand the context $(G, \mathcal{A}, \in)$ as contextual view of $\mathbb{K}$.

3   Measurability

A basic task in the theory of conceptual scaling is to decide if a given formal context $\mathbb{K}$ is derived from plain scaling (up to isomorphism). More precisely, one would like to decide whether $\mathbb{K}$ can be derived using a set of given scales, e.g., from interordinal scaling.

A key result here is Proposition 122 of the FCA book [3]. It answers the following question: Given a formal context $\mathbb{K}$ and a family of scales $\mathcal{S}$, does there exist some many-valued context $\mathbb{D}$ such that $\mathbb{K}$ is (isomorphic to) the context derived from plain scaling of $\mathbb{D}$ using only scales from $\mathcal{S}$? The proposition states that this is the case if and only if $\mathbb{K}$ is fully $\mathcal{S}$-measurable (cf. Definition 94 in Ganter and Wille [3]), i.e., fully measurable into the *semiproduct* of the scales in $\mathcal{S}$.

Based on this proposition, Theorem 55 of the FCA book also gives some simple characterizations for measurability, one of them concerning interordinal scaling:

Theorem 1 (Theorem 55 [3]). *A finite formal context is derivable from interordinal scaling iff it is atomistic (i.e., $\{g\}' \subseteq \{h\}' \implies \{g\}' = \{h\}'$) and the complement of every attribute extent is an extent.*

While this characterization may seem very restrictive there are potential applications machine learning. In Hanika and Hirth [5] the authors discuss how Decision Trees or Random Forest classifiers can be conceptually understood. Similar investigations have been made for latent representations of neural networks [10]. The nature of the decision process suggests an interordinal scaling on the data set. They study several conceptual views based on interordinal scalings, asking if they can be used to explain and interpret tree based classifiers. One of these possibilities they call the **interordinal predicate view** of a set of objects G with respect to a decision tree classifier $\mathcal{T}$. We refer to Hanika and Hirth [5] for a formal definition and a thorough investigation of how to analyze ensembles of decision tree classifiers via views.

Proposition 2. *The interordinal predicate view $\mathbb{I}_{\mathcal{P}}(G, \mathcal{T})$ is derivable from interordinal scaling.*

Proof. The interordinal predicate view context is atomistic, as shown in Proposition 1 of Hanika and Hirth [5]. For every attribute $P \in \mathcal{T}(\mathcal{P})$ we find that $\{P\}' \in \mathrm{Ext}(\mathbb{I}_{\mathcal{P}}(G, \mathcal{T}))$. We have to show that $G \setminus \{P\}' \in \mathrm{Ext}(\mathbb{I}_{\mathcal{P}}(G, \mathcal{T}))$. This is true since we know that for any predicate P that $\{P\}' = \{g \in G \mid g \models P\}$ by definition and therefore $G \setminus \{P\}' \overset{\star}{=} \{g \in G \mid g \not\models P\} = \{\neg P\}'$, which is an extent. The equality $(\star)$ follows directly from the fact that the many valued context $\mathbb{D}$ is complete. Thus, by Sect. 3 the proposition holds. □

Other characterizations (also in Theorem 55 of the FCA book [3]) show that while every context is fully ordinally measurable, a context is fully nominally measurable iff it is atomistic. Thus every fully interordinally measurable context is also fully nominally measurable. In addition to that, the scale families of contranominal, dichtomic and interordinal scales are equally expressive.

While equally expressive, a natural quantity in which these families and scaling in general differs in how complex the scaling is in terms of the size the many-valued context $\mathbb{D}$. This is expressed by the *scaling dimension*, to be introduced in the following definition.

Definition 4 (Scaling Dimension). *Let* $\mathbb{K} := (G, M, I)$ *be a formal context and let* S *be a family of scales. The* **scaling dimension** *of* $\mathbb{K}$ *with respect to* S *is the smallest number* d *such that there exists a many-valued context* $\mathbb{D} := (G, M_{\mathbb{D}}, W_{\mathbb{D}}, I_{\mathbb{D}})$ *with* $|M_{\mathbb{D}}| = d$, *such that* $\mathbb{K}$ *has the same extents as the context derived from* $\mathbb{D}$ *when only scales from* S *are used. If no such scaling exists, the dimension remains undefined.*

The so-defined dimension is related to the feature compression problem. Even when it is known that $\mathbb{K}$ can be derived from a particular data table, it may be that there is another, much simpler table from which one can also derive $\mathbb{K}$ (cf. Figure 1 for an example).

To prove properties of the scaling dimension, we need results about scale measures which were published elsewhere [3,7]. The following lemma follows from Propositions 120 and 122 from Ganter and Wille [3].

Lemma 1. *Let* $\mathbb{D}$ *be a complete many-valued context. For every many-valued attribute* m *of* $\mathbb{D}$ *let* $\mathbb{S}_m := (G_m, M_m, I_m)$ *be a scale for attribute* m, *i.e., with* $W(m) \subseteq G_m$. *Furthermore, let* $\mathbb{K}$ *be the formal context derived from* $\mathbb{D}$ *through plain scaling with the scales* $(\mathbb{S}_m \mid m \in M)$. *Recall that* $\mathbb{K}$ *and* $\mathbb{D}$ *have the same object set* G, *and that for every* $m \in M$ *the mapping* $\sigma_m : G \rightarrow W(m)$, *defined by* $g \mapsto m(g)$, *is a scale measure from* $\mathbb{K}$ *to* $\mathbb{S}_m$.

Then the extents of $\mathbb{K}$ *are exactly the intersections of preimages of extents of the scales* $\mathbb{S}_m$.

A first result on the scaling dimension is easily obtained for the case of ordinal scaling. It was already mentioned that every formal context is fully ordinally measurable, which means that every context is (up to isomorphism) derivable from a many-valued context $\mathbb{D}$ through plain ordinal scaling. But how large must this $\mathbb{D}$ be, how many many-valued attributes are needed? The next proposition gives the answer. For simplicity, we restrict to the finite case.

Proposition 3. *The ordinal scaling dimension of a finite formal context* $\mathbb{K}$ *equals the width of the ordered set of infimum-irreducible concepts.*

Proof. The width is equal to the smallest number of chains C_i covering the $\subseteq$-ordered set of irreducible attribute extents. From these chains we can construct a many-valued context $\mathbb{D}$ with one many-valued attribute m_i per chain C_i. The values of m_i are the elements of the chain C_i, and the order of that chain is understood as an ordinal prescaling. The derived context by means of ordinal scaling has exactly the set of all intersections of chain extents as extents (Proposition 120 [3]), i.e., the set of all $\bigcap \mathcal{A}$ where $\mathcal{A} \subseteq C_1 \times \cdots \times C_w$. Those are exactly the extents of $\mathbb{K}$. This implies that the scaling dimension is less or equal to the width.

But the converse inequality holds as well. Suppose $\mathbb{K}$ has ordinal scaling dimension w. Then by Lemma 1 every extent of $\mathbb{K}$ is the intersection of preimages of extents of the individual scales. For $\cap$-irreducible extents this means that they must each be a preimage of an extent from one of the scales. Incomparable extents cannot come from the same (ordinal) scale, and thus the scaling must use at least w many ordinal scales. ☐

As a proposition we obtain that the ordinal scaling dimension must be at least as large as the order dimension:

Proposition 4 (Ordinal Scaling Dimension and Order Dimension).
The order dimension of the concept lattice $\mathfrak{B}(\mathbb{K})$ is a lower bound for the ordinal scaling dimension of $\mathbb{K}$.

Proof. It is well known that the order dimension of $\mathfrak{B}(\mathbb{K})$ equals the Ferrers dimension of $\mathbb{K}$ [3, Theorem 46], which remains the same when $\mathbb{K}$ is the standard context. The Ferrers relation is the smallest number of staircase-shaped relations to fill the complement of the incidence relation of $\mathbb{K}$.

For a context with ordinal scaling dimension equal to w we can conclude that the (irreducible) attributes can be partitioned into w parts, one for each chain, such that for each part the incidence is staircase-shaped, and so are the non-incidences. Thus we can derive w Ferrers relations to fill all non-incidences.   □

A simple example that order dimension and ordinal scaling dimension are not necessarily the same is provided by any context with the following incidence:

Its Ferrers dimension is two, but there are three pairwise incomparable irreducible attributes, which forces its ordinal scaling dimension to be three.

A more challenging problem is to determine the interordinal scaling dimension of a context $\mathbb{K}$. We investigate this with the help of the following definition.

Definition 5. *An **extent ladder** of $\mathbb{K}$ is a set $\mathcal{R} \subseteq \operatorname{Ext}(\mathbb{K})$ of nonempty extents that satisfies:*

i) the ordered set $(\mathcal{R}, \subseteq)$ has width ≤ 2, i.e., $\mathcal{R}$ does not contain three mutually incomparable extents, and
ii) $\mathcal{R}$ is closed under complementation, i.e., when $A \in \mathcal{R}$, then also $G \setminus A \in \mathcal{R}$.

Note that a finite (and nonempty) extent ladder is the disjoint union of two chains of equal cardinality, for the following reason: Consider a minimal extent E in the ladder. Any other extent must either contain E or be contained in the complement of E, because otherwise there would be three incomparable extents. The extents containing E must form a chain, and so do their complements, which are all contained in the complement of E.

Theorem 2 (Interordinal Scaling Dimension).   *The interordinal scaling dimension of a finite formal context $\mathbb{K}$, if it exists, is equal to the smallest number of extent ladders, the union of which contains all meet-irreducible extents of $\mathbb{K}$.*

Proof. Let $\mathbb{K}$ be a formal context with interordinal scaling dimension d. W.l.o.g. we may assume that $\mathbb{K}$ was derived by plain interordinal scaling from a many-valued context $\mathbb{D}$ with d many-valued attributes. We have to show that the

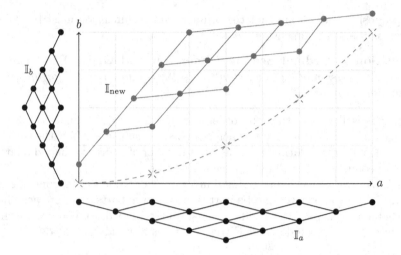

Fig. 1. This figure displays the a and b feature of five data points and their respective interordinal scales $\mathbb{I}_a$ and $\mathbb{I}_b$ (black). The interordinal scaling dimension of this data set is one and the respective reduced interodinal scale $\mathbb{I}_{\text{new}}$ is depicted in red. The reduction would then remove the a and b data column and substitute it for a new column given by $\mathbb{I}_{\text{new}}$.

irreducible attribute extents of $\mathbb{K}$ can be covered by d extent ladders, but not by fewer.

To show that d extent ladders suffice, note that the extents of an interordinal scale form a ladder, and so do their preimages under a scale measure. Thus Lemma 1 provides an extent ladder for each of the d scales, and every extent is an intersection of those. Meet-irreducible extents cannot be obtained from a proper intersection and therefore must all be contained in one of these ladders.

For the converse assume that $\mathbb{K}$ contains l ladders covering all meet-irreducible extents. From each such ladder $\mathcal{R}_i$ we define a formal context $\mathbb{R}_i$, the attribute extents of which are precisely the extents of that ladder, and note that this context is an interordinal scale (up to clarification). Define a many-valued context with l many-valued attributes m_i. The attribute values of m_i are the minimal non-empty intersections of ladder extents, and the incidence is declared by the rule that an object g has the value V for the attribute m_i if $g \in V$. The formal context derived from this many-valued context by plain interordinal scaling with the scales $\mathbb{R}_i$ has the same meet-irreducible extents as $\mathbb{K}$, and therefore the same interordinal scaling dimension. Thus $l \geq d$. $\square$

Proposition 5. *Let w denote the width of the ordered set of meet-irreducible extents of the formal context $\mathbb{K}$. The interordinal scaling dimension of $\mathbb{K}$, if defined, is bounded below by $w/2$ and bounded above by w.*

Proof. An extent ladder consists of two chains, and w is the smallest number of chains covering the meet-irreducible extents. So at least $w/2$ ladders are required.

Conversely from any covering of the irreducible extents by w chains a family of w ladders is obtained by taking each of these chains together with the complements of its extents.   □

The last inequality $\mathrm{OSD}(\mathbb{K}) \leq 2 \cdot \mathrm{ISD}(\mathbb{K})$ results from using each two chains of the extent ladders as ordinal scales, where $\mathrm{OSD}(\mathbb{K})$ denotes the ordinal scaling dimension and $\mathrm{ISD}(\mathbb{K})$ the interordinal scaling dimension of $\mathbb{K}$. This results in an upper bound for the ordinal scaling dimension and a lower bound for the interordinal scaling dimension. A context where $\mathrm{OSD}(\mathbb{K}) \neq 2 \cdot \mathrm{ISD}(\mathbb{K})$ is depicted in the next section in Fig. 4.

Another inequality that can be found in terms of many-valued contexts. For a many-valued context $\mathbb{D}$ and its ordinal scaled context $\mathbb{O}(\mathbb{D})$ and interordinal scaled context $\mathbb{I}(\mathbb{D})$ is the ISD of $\mathbb{I}(\mathbb{D})$ in general not equal to the OSD of $\mathbb{O}(\mathbb{D})$. Consider for this the counter example given in Fig. 2. The depicted many-valued context has two ordinally pre-scaled attributes that form equivalent interordinal scales.

$\mathbb{D}$	m_1	m_2
g_1	1	d
g_2	2	c
g_3	3	b
g_4	4	a

Fig. 2. Example many-valued context where the attribute values are ordinally pre-scaled by $1 < 2 < 3 < 4$ and $a < b < c < d$. The interordinal scaling dimension of $\mathbb{I}(\mathbb{D})$ is one and the ordinal scaling dimension of $\mathbb{O}(\mathbb{D})$ is two.

4   Small Case Study

To consolidate the understanding of the notions and statements on the (interordinal) scaling dimension we provide an explanation based on a small case example taken from the *drive concepts* [3] data set. This data set is a many-valued context consisting of five objects, which characterize different ways of arranging the engine and drive chain of a car, and seven many-valued attributes that measure

	De +	C l	M +	E +	R +	Dl +
Conventional			×	×	×	×
All-Wheel	×			×	×	×
Mid-Wheel	×	×			×	×
Rear-Wheel	×	×	×			×
Front-Wheel	×	×	×	×	×	

Fig. 3. The standard context of the drive concepts lattice, cf. Figure 1.13 in Ganter and Wille [3].

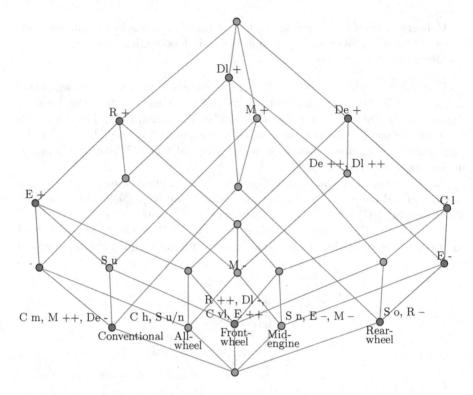

Fig. 4. Concept lattice for the context of drive concepts (cf. Figure 1.13 and 1.14 in Ganter and Wille [3]). The extent ladders indicating the three interordinal scales are highlighted in color. The ordinal scaling dimension as well as order dimension of this context is four.

quality aspects for the driver, e.g., *economy of space*. The data set is accompanied by a scaling that consists of a mixture of bi-ordinal scalings of the quality (attribute) features, e.g., *good* < *excellent* and *very poor* < *poor*, and a nominal scaling for categorical features, e.g., for the *steering behavior*. The concept lattice of the scaled context consists of twenty-four formal concepts and is depicted in Fig. 4.

First we observe that the concept lattice of the example meets the requirements to be derivable from interordinal scaling (Sect. 3). All objects are annotated to the atom concepts and the complement of every attribute extent is an extent as well. The interordinal scaling dimension of the scaled *drive concept* context is three which is much smaller than the original seven many-valued attributes. Using the extent ladder characterization provided in Theorem 2 we highlighted three extent ladders in color in the concept lattice diagram (see Fig. 4). The first and largest extent ladder (highlighted in red) can be inferred from the outer most concepts and covers sixteen out of twenty-four concepts. The remaining two extent ladders have only two elements and are of dichotomic scale.

5   Discussion and Future Work

The presented results on the scaling dimension have a number of interfaces and correspondences to classical data science methods. A natural link to investigate would be comparing the scaling dimension with standard correlation measures. Two features that correlate prefectly, e.g., Fig. 1, induce an equivalent conceptual scaling on the data. An analog of the scaling dimension in this setting would be the smallest number of independent features. Or, less strict, the smallest number of features such that these features do not correlate more than some parameter. This obvious similarity of both methods is breached by a key advantage of our approach. In contrast to correlation measures, our method relies solely on ordinal properties [14] and does not require the introduction of measurements for distance or ratios.

Proposition 4 has already shown that there is a relationship between an aspect of the scaling dimension of a formal context and the order dimension of its concept lattice. The assumption that further such relationships may exist is therefore reasonable. Yet, a thorough investigation of these relationships is an extensive research program in its own right and therefore cannot be addressed within the scope of this paper. An investigation on how the scaling dimension relates to other measures of dimension within the realm of FCA [9,15] is therefore deemed future work.

Due to novel insights into the computational tractability of recognizing scale-measures [7] (that is in preparation and will be made public later this year) we have little hope that the scaling dimension and interordinal scaling dimension can be decided in polynomial time. Despite that, efficient algorithms for real-world data that compute the scaling dimension and its specific versions, i.e., ordinal, interordinal, nominal, etc., may be developed. In addition to that, so far it is unknown if an approximation of the scaling dimension, e.g., with respect to some degree of conceptual scaling error [8] or bounds, is tractable. If computationally feasible, such an approximation could allow larger data sets to be handled.

Another line of research that can be pursued in future work is how the scaling dimension can be utilized to derive more readable line diagrams. We can envision that diagrams of concept lattices that are composed of fewer scales, i.e., have a lower scaling dimension, are more readable even if they have slightly more concepts. An open problem that needs to be solved here is: for a context $\mathbb{K}$ and $k \in \mathbb{N}$ identify k scales that cover the largest number of concepts from $\mathfrak{B}(\mathbb{K})$ with respect to scale measures.

6   Conclusion

With our work, we contributed towards a deeper understanding of conceptual scaling [2]. In particular, we introduced the notion of pre-scaling to formalize background knowledge on attribute domains, e.g., underlying order relations, that can be used and extended to scales in the process of conceptual scaling. To deal with the complexity of scalings selection or logical compression methods have been proposed to reflect parts of the conceptual structure [7]. Furthermore, we introduced the notions of conceptual and contextual views to characterize these methods and provided a first formal definition.

We extended the realm of conceptual measurability [4] by the scaling dimension, i.e., the least number of attributes needed to derive a context by the means of plain scaling. This notion does not only provide insight towards the complexity of an underlying scaling but can also be applied for many-valued feature compression. For the identification of the scaling dimension, we provided characterizations for the ordinal and interordinal scaling dimension in terms of structural properties of the concept lattice. These employ chains of meet-irreducible extents and newly introduced extent ladders. We demonstrated their applicability based on the drive concepts data set and highlighted the identified extent ladders and chains in the concept lattice diagram. Our analysis showed that while the many-valued context consists of seven many-valued attributes an equivalent scaling can be derived from three interordinally scaled or four ordinal scaled many-valued attributes.

In addition to the structural characterizations of the scaling dimensions, we provided bounds for the interordinal and ordinal scaling dimension. In detail, we showed upper and lower bounds in terms of the width and the order dimension of the concept lattice. This result shows in particular how far-reaching and therefore necessary a future in-depth investigation of the scaling dimension is.

References

1. Cimiano, P., Hotho, A., Stumme, G., Tane, J.: Conceptual knowledge processing with formal concept analysis and ontologies. In: Eklund, P. (ed.) ICFCA 2004. LNCS (LNAI), vol. 2961, pp. 189–207. Springer, Heidelberg (2004). https://doi.org/10.1007/978-3-540-24651-0_18
2. Ganter, B., Wille, R.: Conceptual scaling. In: Roberts, F. (eds.) Applications of Combinatorics and Graph Theory to the Biological and Social Sciences. The IMA Volumes in Mathematics and Its Applications, vol. 17, pp. 139–167. Springer, New York (1989). https://doi.org/10.1007/978-1-4684-6381-1_6
3. Ganter, B., Wille, R.: Formal Concept Analysis: Mathematical Foundations. Springer, Berlin, pp. x+284 (1999)
4. Ganter, B., Stahl, J., Wille, R.: Conceptual measurement and many-valued contexts. In: Gaul, W., Schader, M. (eds.) Classification as a tool of research, pp. 169–176. North-Holland, Amsterdam (1986)
5. Hanika, T., Hirth, J.: Conceptual views on tree ensemble classifiers. Int. J. Approximate Reasoning **159**, 108930 (2023). https://doi.org/10.1016/j.ijar.2023.108930
6. Hanika , T., Hirth, J.: Conexp-Clj - a research tool for FCA. In: Cristea, D., et al. (ed) ICFCA (Suppl.), vol. 2378. CEUR-WS.org, pp. 70–75 (2019)
7. Hanika, T., Hirth, J.: On the lattice of conceptual measurements. Inf. Sci. **613**, 453–468 (2022). https://doi.org/10.1016/j.ins.2022.09.005
8. Hanika, T., Hirth, J.: Quantifying the conceptual error in dimensionality reduction. In: Braun, T., Gehrke, M., Hanika, T., Hernandez, N. (eds.) ICCS 2021. LNCS (LNAI), vol. 12879, pp. 105–118. Springer, Cham (2021). https://doi.org/10.1007/978-3-030-86982-3_8

9. Hanika, T., Schneider, F.M., Stumme, G.: Intrinsic dimension of geometric data sets. In: Tohoku Mathematical Journal (2018)
10. Hirth, J., Hanika, T.: Formal conceptual views in neural networks (2022). https://arxiv.org/abs/2209.13517
11. Kaytoue, M., Duplessis, S., Kuznetsov, S.O., Napoli, A.: Two FCA-based methods for mining gene expression data. In: Ferré, S., Rudolph, S. (eds.) ICFCA 2009. LNCS (LNAI), vol. 5548, pp. 251–266. Springer, Heidelberg (2009). https://doi.org/10.1007/978-3-642-01815-2_19
12. Prediger, S., Stumme, G.: Theory-driven logical scaling. In: Proceedings of 6th International Workshop Knowledge Representation Meets Databases (KRDB 1999). Ed. by E. F. et al. vol. CEUR Workshop Proc. 21. Also. In: P. Lambrix et al. (eds.): Proceedings of International WS on Description Logics (DL 1999). CEUR Workshop Proc. vol. 22, 1999 (1999)
13. Prediger, S.: Logical scaling in formal concept analysis. In: Lukose, D., Delugach, H., Keeler, M., Searle, L., Sowa, J. (eds.) ICCS-ConceptStruct 1997. LNCS, vol. 1257, pp. 332–341. Springer, Heidelberg (1997). https://doi.org/10.1007/BFb0027881
14. Stevens, S.S. On the theory of scales of measurement. Science 103(2684), 677–680 (1946). ISSN: 0036–8075
15. Tatti, N., et al.: What is the dimension of your binary data? In: Sixth International Conference on Data Mining (ICDM 2006), pp. 603–612 (2006)

Three Views on Dependency Covers from an FCA Perspective

Jaume Baixeries[1]([envelope]) [iD], Victor Codocedo[2] [iD], Mehdi Kaytoue[3] [iD],
and Amedeo Napoli[4] [iD]

[1] Universitat Politècnica de Catalunya, Barcelona, Catalonia
jbaixer@cs.upc.edu
[2] Instituto para la Resiliencia ante Desastres, Santiago, Chile
[3] Université de Lyon. CNRS, INSA-Lyon, LIRIS, Lyon, France
[4] Université de Lorraine, CNRS, LORIA, Nancy, France
amedeo.napoli@loria.fr

Abstract. Implications in Formal Concept Analysis (FCA), Horn clauses in Logic, and Functional Dependencies (FDs) in the Relational Database Model, are very important dependency types in their respective fields. Moreover, they have been proved to be equivalent from a syntactical point of view. Then notions and algorithms related to one dependency type in a field can be reused and applied to another dependency type in the other field. One of these notions is that of *cover*, also known as a *basis*, i.e., a compact representation of a complete set of implications, FDs, or Horn clauses. Although the notion of cover exists in the three fields, the characterization and the related uses of a cover are different. In this paper, we study and compare, from an FCA perspective, the principles on which rely the most important covers in each field. Finally, we discuss some open questions that are of interest in the three fields, and especially to the FCA community.

Keywords: Functional dependencies · Implications · Horn Clauses · Dependency Covers · Closure

1 Introduction and Motivation

A **dependency** is a relation between sets of attributes in a dataset. In this paper, they are represented as $X \rightarrow Y$, where the type of the subsets of attributes X and Y, and the semantics of $\rightarrow$ may vary w.r.t. the context. There are many different kinds of dependencies: complete and comprehensive surveys, from a Relational Database Theory perspective, can be found in [18] and in [13]. Here, we focus on those dependencies that follow the so called Armstrong axioms, this is, reflexivity, augmentation and transitivity, which appear in different fields of computer science: *functional dependencies* (FDs) in the Relational Database Model, *Horn clauses* in logic and logic programming, and *implications* in Formal Concept Analysis.

D. Dürrschnabel and D. López Rodríguez (Eds.): ICFCA 2023, LNAI 13934, pp. 78–94, 2023.
https://doi.org/10.1007/978-3-031-35949-1_6

Functional dependencies [27] are of paramount importance in the Relational Database Model (RDBM), where they are used to express constraints or rules that need to hold in a database, to help the design of a database or to check for errors and inconsistencies. A set of **Horn clauses** [16] is a special case of Boolean functions that are crucial in logic programming [19,23] and artificial intelligence (see [9] for a detailed explanation). **Implications** are at the core of Formal Concept Analysis (FCA) where they are used to model and deduce relevant information that is contained in a formal context [14,15].

Although all three of them appear in different fields, have been applied on different kinds of data and have been used for different purposes, they all share the same axioms, which means that, from a syntactical point of view, they are all equivalent. More specifically, the equivalence between functional dependencies and Horn clauses is presented in [17,25] (see also [9] for a more detailed explanation). The equivalence between functional dependencies and implications is explained in [15] and the equivalence between implications and Horn clauses is explained in Sect. 5.1 in [14] as well as in [8]. These equivalences allow us to talk in a generic way of **Armstrong dependencies** or, simply, **dependencies**.

One of the consequences of this equivalence is the transversality of concepts, problems and algorithms between these three fields. One of the most typical examples is the decision problem of the **logical implication** which consists in, given a set of dependencies Σ and a single dependency σ, to determine whether σ is logically entailed by Σ, that is, $\Sigma \models \sigma$. Entailment means that σ can be derived from Σ by the iterative application of the Armstrong axioms. This problem is named **implication problem** in the RDBM [1,20] and FCA fields, and **deduction** in logic [9]. It is of capital interest in all three fields. In the RDBM it allows to test whether two different sets of functional dependencies are equivalent [13], and it also allows to compute a more succinct set of functional dependencies, which is relevant to assist the design process of databases [4,22]. In logic the deduction problem is used to check whether a logical expression is consistent w.r.t. a knowledge base and to compute the prime implicants of a Horn function [9]. In Formal Concept Analysis this problem is used, for instance, in attribute exploration [14], which consists in creating a data table (context) in agreement w.r.t. a set of attributes and a set of objects, and also for computing the Duquenne-Guigues basis [11].

Roughly speaking, the computation of the logical implication problem $\Sigma \models X \to Y$ is performed by iterating over Σ and applying the Armstrong axioms to infer new dependencies until a positive or negative answer is found. However, this problem can be reduced to the computation of the **closure** of X with respect to Σ (closure$_\Sigma(X)$). This closure returns the largest set such that $\Sigma \models X \to$ closure$_\Sigma(X)$ holds. Therefore, the implication problem $\Sigma \models X \to Y$ boils down to testing whether $Y \subseteq$ closure$_\Sigma(X)$.

As an example of transversality, the algorithm that computes closure$_\Sigma(X)$ appears in most of the main database textbooks, where it is called **closure** [1, 20,27], and also in logic, where it is called **forward chaining** [9], while in FCA the same algorithm that first appeared in the RDBM is discussed and reused in [14]. An improved version of Closure is Linclosure [4]. Both Closure and

Linclosure have two parameters: a set of attributes X and a set of dependencies Σ. The performance of both algorithms is first determined by their worst-case complexity: quadratic in the case of Closure, linear in the case of Linclosure, always with respect to the size of Σ, although the nature and the size of Σ greatly affect the performance of both algorithms outside of the worst case.

Another "transversal notion" which is present in all three fields is the notion of **cover**. In general terms, it is not suitable to handle the set Σ of all dependencies that hold, because of its potential large size, but rather a subset of Σ that contains the same information and that is significantly smaller in size. By *"containing the same information"* we mean that this subset may generate, thanks to the application of the Armstrong axioms, the complete set Σ. This compact and representative subset is called "cover" in the RDBM, "basis" in FCA and "set of prime implicants" in logic. Moreover, each field has defined and used a different kind of cover. While in the RDBM this base is the Canonical-Direct Basis or the Minimal Cover, the cover of choice in FCA is the so-called Duquenne-Guigues basis.

Both the implication problem and the problem of computing a cover are related: the implication problem is used in the algorithm **Canonical Basis** (Algorithm 16, page 103 in [14]) to compute the Duquenne-Guigues basis, and it is also used in the algorithm **Direct** (Sect. 5.4, Chap. 5 in [20]) which is used to compute the Minimal Cover. Again, the transversality of the Armstrong dependencies appears in a general concept (computing a cover) but in different forms (Duquenne-Guigues basis and Minimal Cover).

The purpose of this study is to present in a single paper the main different covers used in the RDBM, in Logic, and in FCA, and to discuss two closely related questions: (i) the computation of the closure of a set of attributes w.r.t. a set of Armstrong dependencies, (ii) the representation of such a set of the dependencies in a cover. To do so, we first present both Closure and Linclosure, debate their complexity and the different factors that may affect their performance. Afterwards, we review three different main covers that appeared in the literature. To the best of our knowledge, this is the first time that such a comprehensive study is proposed, where the connections between the relational database model, Logic, and FCA are examined from the implication problem perspective.

The paper is organized as follows. Section 2 presents the main aspects of the implication problem in the RDBM, Logic, and FCA. Then Sect. 3 provides the necessary definitions needed in the paper. Section 4 makes precise the Closure and Linclosure algorithms, while Sect. 5 includes a detailed comparison of the main covers. Finally, Sect. 6 concludes the paper and proposes an extensive discussion about these different and important covers.

2   The Relevance of the Implication Problem

In this section we briefly explain why the implication problem, that is, to check whether a dependency σ can be derived from a set of dependencies Σ by the iterative application of the Armstrong axioms is of importance in RDBM, logic, and FCA. We start by quoting the survey [13] which comes from the Relational Database field:

> *Most of the papers in dependency theory deal exclusively with various aspects of the implication problem, that is, the problem of deciding for a given set of dependencies Σ and a dependency σ whether Σ logically implies σ. The reason for the prominence of this problem is that an algorithm for testing implication of dependencies enables us to test whether two given sets of dependencies are equivalent or whether a given set of dependencies is redundant. A solution for the last two problems seems a significant step towards automated database schema design, which some researchers see as the ultimate goal for research in dependency theory.*

In what way are those two problems *a significant step towards automated database schema design* and what is the reason of the *prominence of* the implication problem? Functional dependencies are used to design relational schemes (see for example the extensive Chap. 12.2 *Nonloss decomposition and functional dependencies* in [10]). However, *some of the FDs that are used to describe the data base may be redundant in the sense that they are derivable from the others.* These (redundant) FDs used to assist this method of schema design *will induce redundancy in the relational schema,* which is to be avoided since *redundancy in relations creates serious problems in maintaining a consistent data base.* Hence, one needs to be able to compute a *covering* (here we will use the term *cover*) of a set of dependencies, this is, a non-redundant set of FDs that yields the same closure with respect to the axioms for FDs. And, *the problem of finding a covering reduces to computing the predicate $\sigma \in \Sigma^+$, that is, $\Sigma \models \sigma$* (all portions of text in *italic* are citations from [6]).

Summing up, the implication problem for FDs is of importance because it solves the problem of computing a cover of a set of FDs which, in turn, prevents the propagation of redundancy in the design of a database scheme using functional dependencies.

In FCA, the basic data structure is the binary context which can have two related representations, namely the concept lattice and the basis of implications. The latter is the Duquenne-Guigues basis which is unique, minimal, and non redundant (see 5). There is an equivalence between these three views of an initial dataset. Moreover, one can be also interested in the so-called equivalence classes which are associated with one closed set and possibly several generators of different types [21]. A typical implication is related to any equivalence class which is of the form $X \rightarrow Y$ where Y is a closed set and closure$(X) = Y$. Then, a minimal basis has all its importance since it provides a summary of the dataset under study with a minimum number of elements. In particular, the number of implications in say the Duquenne-Guigues basis is much smaller than the number of concepts, that is, a substantial number of implications can be inferred from this minimum basis. This is also of importance when the problem of construction or reconstruction is considered, that is, starting with a set of implications, design the whole related dataset. This problem is present in the RDBM, in Logic and especially in the design of a theory or of an ontology, and as well in FCA with the attribute exploration process [14].

3   Definitions

In this section we introduce the definitions used in this paper. We do not provide the references for all of them because they can be found in all the textbooks and papers related to the RDBM, Logic and FCA.

As explained in the introduction, implications [15], functional dependencies [20] and Horn clauses [16] are dependencies between sets of attributes, which are equivalent from a syntactical point of view, since they are in agreement with the Armstrong axioms.

Definition 1. *Given set of attributes $\mathcal{U}$, for any $X, Y, Z \subseteq \mathcal{U}$, the Armstrong axioms are:*

1. **Reflexivity:** *If $Y \subseteq X$, then $X \to Y$ holds.*
2. **Augmentation.** *If $X \to Y$ holds, then $XZ \to YZ$ holds.*
3. **Transitivity.** *If $X \to Y$ and $Y \to Z$ hold, then $X \to Z$ holds.*

When we write that a dependency $X \to Y$ **holds,** we mean all the instances in which this dependency is valid or true. Therefore, the sentence *"If $X \to Y$ holds, then $XZ \to YZ$ holds"* can be rephrased as *"In any instance in which $X \to Y$ is valid, the dependency $XZ \to YZ$ is valid as well".*

The Armstrong axioms allow us to define the closure of a set of dependencies as the iterative application of these axioms over a set of dependencies.

Definition 2. *Σ^+ denotes the closure of a set of dependencies Σ, and can be constructed thanks to the iterative application of the Armstrong axioms over Σ. This iterative application terminates when no new dependency can be added, and it is finite. Therefore, Σ^+ contains the largest set of dependencies that hold in all instances in which all the dependencies in Σ hold.*

The closure of a set of dependencies induces the definition of the cover of such a set of dependencies.

Definition 3. *The **cover** or **basis** of a set of dependencies Σ is any set Σ' such that $\Sigma'^+ = \Sigma^+$.*

We define now the concept of a *closure* of a set of attributes $X \subseteq \mathcal{U}$ with respect to a set of implications Σ.

Definition 4. *The closure of X with respect to a set of dependencies Σ is*

$$\text{closure}_\Sigma(X) = \{\, Y \mid X \to Y \in \Sigma^+ \,\}$$

that is, $\text{closure}_\Sigma(X)$ is the largest set of attributes Y such that $X \to Y$ can be derived by the iterative application of the Armstrong axioms over the set Σ.

This closure operation returns the largest set of attributes such that $\Sigma \models X \to \text{closure}_\Sigma(X)$. Therefore, the implication problem $\Sigma \models X \to Y$ boils down to testing whether $Y \subseteq \text{closure}_\Sigma(X)$.

4   Algorithms to Compute the Closure of a Set of Attributes

Below we review two most well-known algorithms to compute closure$_\Sigma$.

4.1   The Closure Algorithm

The first algorithm Closure appears in most of the main database textbooks, e.g., [1,20,27], and also in logic, where it is called **forward chaining** [9]. In Formal Concept Analysis the same algorithm that first appeared in the RDBM is reused as for example in [14].

Function Closure(X, Σ)

> **Input** : A set of attributes $X \subseteq \mathcal{U}$ and a set of implications Σ
> **Output:** closure$_\Sigma(X)$

```
1  stable ← false
2  while not stable do                              // Outer loop
3  │   stable ← true
4  │   forall A → B ∈ Σ do                          // Inner loop
5  │   │   if A ⊆ X then
6  │   │   │   X ← X ∪ B
7  │   │   │   stable ← false
8  │   │   │   Σ ← Σ \ { A → B }
9  │   │   end
10 │   end
11 end
12 return X
```

In Table 1 we list different complexities of Closure as they are given in a classic database textbook [20], a FCA textbook [14] and the pioneer paper of Linclosure [4][1]. In general terms, the complexity of Closure depends on its two loops which are marked in the code as outer and inner loops. This algorithm iterates in the outer loop as long as there is a change in the computation of closure$_\Sigma(X)$. This means that in the worst case, the closure may be incremented by only one single attribute at each iteration of the outer loop, which implies that the outer loop is of order $\mathcal{O}(|\mathcal{U}|)$. Regarding the inner loop, it necessarily iterates over all the dependencies that are in Σ, that is, it is of order $\mathcal{O}(|\Sigma|)$. Then the total number of iterations, in the worst case, is of order $\mathcal{O}(|\mathcal{U}| \times |\Sigma|)$. Since in some cases it may happen that $|\Sigma| = |\mathcal{U}|$, the complexity of this algorithm can be, in the worst case, of order $\mathcal{O}(|\Sigma|^2)$.

---

[1] Other relevant textbooks on RDBM line [1,27] provide the same complexity analysis as in Table 1.

However, there are two extra comments about this complexity analysis. First, we have stated that the inner loop (line 4) iterates over all the dependencies in Σ, but all the dependencies that have been used to compute closure$_\Sigma(X)$ are deleted in line 8. But even if this removal is performed, the number of iterations is still of order $|\Sigma|$, since in the worst case, we may remove only one single dependency from Σ at each iteration of the inner loop. Also, in line 5, there is a subset containment check of order $\mathcal{O}(|\mathcal{U}|)$ that is performed as many times as there are iterations of the inner loop. This fact, which is only considered in [20], induces a complexity of order $\mathcal{O}(|\mathcal{U}| \times |\Sigma|^2)$. As we can observe in Table 1, the consensus is that the complexity of Closure is of order $\mathcal{O}(|\Sigma|^2)$ (in the worst case scenario $|\mathcal{U}| = |\Sigma|$). However, in cases when $|\mathcal{U}| << |\Sigma|$, the complexity of Closure becomes of order $\mathcal{O}(|\mathcal{U}| \times |\Sigma|)$. One of these examples can be found in Sect. 3.2.3 in [14], where $|\Sigma|$ is exponential w.r.t. $|\mathcal{U}|^2$.

In any case, Closure is considered a **quadratic** time algorithm with respect to the number of dependencies.

Table 1. Complexity of the Closure algorithm according to different authors.

Ref	Inner	Outer	Total
Maier [20]	$|\mathcal{U}| \times |\Sigma|$	$|\Sigma|$	$|\mathcal{U}| \times |\Sigma|^2$
Ganter-Obiedkov [14]	$|\Sigma|$	$min(|\mathcal{U}|, |\Sigma|)$	$min(|\mathcal{U}| \times |\Sigma|, |\Sigma|^2)$
Beeri-Bernstein [4]	$|\Sigma|$	$|\mathcal{U}|$	$|\mathcal{U}| \times |\Sigma|$

4.2   The Linclosure Algorithm

An improved version of Closure is the Linclosure algorithm [4]. In particular, Linclosure improves Closure by annotating in the preparation part the dependencies which do not need to be checked to compute closure$_\Sigma(X)$. It should be noticed that the "chain forward algorithm" of Marcel Wild [28] is in fact an improvement of Linclosure which is not covered in this paper.

Linclosure consists of two parts: a `preparation` part in which the necessary data structures are computed, and the `computation` part in which the computation of closure$_\Sigma(X)$ is performed. In the first part two data structures are designed: one structure contains, for each attribute say Z, a pointer to all the dependencies $X \to Y$ such that Z appears in the left-hand side X. The second structure is such that for each dependency $X \to Y$ there is a counter that annotates the number of attributes of its left-hand side X which have already been visited while computing closure$_\Sigma(X)$. The objective of these two structures is to only visit the dependencies that are necessary to compute the closure, and to ignore those that are not useful to compute it. Regarding the complexity of

---

[2] "*the number of pseudo-closed sets* [Σ] *can be exponential in comparison to the size of the base set* [$\mathcal{U}$]".

Function LINCLOSURE(X, Σ)

Input   : A set of attributes $X \subseteq \mathcal{U}$ and a set of implications Σ
Output: closure$_\Sigma(X)$

```
1  forall A → B ∈ Σ do                                    // Preparation
2  │   count[A → B] ← | A |
3  │   if | A |= 0 then
4  │   │   X ← X ∪ B
5  │   end
6  │   forall a ∈ A do
7  │   │   list[a] ← list[a] ∪ { A → B }
8  │   end
9  end
10 update ← X
```

```
11 while update ≠ ∅ do                                    // Outer loop
12 │   choose m ∈ update
13 │   update ← update \ { m }
14 │   forall A → B ∈ list[m] do                          // Inner loop
15 │   │   count[A → B] ← count[A → B] − 1
16 │   │   if count[A → B] = 0 then
17 │   │   │   add ← B \ X
18 │   │   │   X ← X ∪ add
19 │   │   │   update ← update ∪ add
20 │   │   end
21 │   end
22 end
23 return X
```

Linclosure, there is a general consensus that this algorithm is of order $\mathcal{O}(|\Sigma|)$ (here we do not discuss the complexity of the preparation part, which is assumed to be of the same complexity as the rest of the algorithm). One explanation of this fact appears in the pioneering paper [4], page 47 in the second paragraph[3]:

> For each attribute in [update], the [outer] loop follows a constant number of steps for each occurrence of that attribute on the left side of an FD in Σ. Similarly, each right side of an FD in Σ is visited at most once in [the outer loop]. Thus [the outer loop] is also $\mathcal{O}(|\Sigma|)$ as is the entire Algorithm[4].

Two comments should be made here. The first one is that the sentence *"each right side of an FD in Σ is visited at most once"* also applies to Closure (lines 6

---

[3] We adapt this paragraph to fit names in Algorithm Linclosure.

[4] *Previously, the authors have concluded that the complexity of the preparation part is of order $\mathcal{O}(|\Sigma|)$, as well as the second part, hence "is also $\mathcal{O}(|\Sigma|)$ the entire Algorithm".*

and 7) since a dependency is removed once it has been used (line 8 of Closure). The second one is that *"For each attribute in [update], the [outer] loop follows a constant number of steps for each occurrence of that attribute on the left side of an FD in Σ"* tells us that for each attribute in *update* (which is the variable that contains the attributes that still need to be processed in the outer loop) there is a **constant** number of steps that are performed in each left hand side **of each dependency that contains that attribute** (in its left-hand side). But this number of dependencies is of order $\mathcal{O}(|\Sigma|)$.

From our understanding, this is relevant since the total number of iterations that are performed in Linclosure is the same as the number of times that the algorithm executes lines 15 and 16, and this number may be, in the worst case, when **all** the dependencies are decremented as many times as the number of attributes in their left-hand side, that is, of order $\mathcal{O}(|\mathcal{U}| \times |\Sigma|)$. And here we have a caveat: the complexity of both Closure and Linclosure is given with respect to the **size of the input**, that is, the size of Σ. However, in some cases this size is interpreted as the number of dependencies that are in Σ, that is, $|\Sigma|$, and in some other cases, it is interpreted as the size in number of symbols, that is, $|\mathcal{U}| \times |\Sigma|$. For example, in [20], the complexity of Linclosure is of order $\mathcal{O}(n)$ for an input length n, where $n = |\mathcal{U}| \times |\Sigma|$, whereas in [4] this complexity, is of order $\mathcal{O}(|\Sigma|)$ as we have stated.

4.3 Experimental Evaluation of Closure and Linclosure

As mentioned above, the general consensus is that the complexity order of Closure is quadratic in the worst case, whereas Linclosure is assumed to be of linear cost. But several simple examples (see [3], for example) show that the performance of both algorithms is sensitive to different factors, e.g., the size of Σ or the order in which Σ is explored. Moreover, the expected performance of Closure and Linclosure did not fully match their empirical evaluation in [3]. To be more precise, the experiments were not fully conclusive in supporting the theoretical complexity analysis, since in some scenarios Closure unexpectedly outperformed Linclosure. In other words, Linclosure did not clearly outperform Closure in all scenarios as expected.

Actually, we need to take into account that closure$_\Sigma$ has two parameters, namely a set of dependencies Σ and a set of attributes X, and the complexity of the algorithms is given with respect to the size of both sets. In this case, the only parameter that admits some degree of variation is Σ. Therefore, there is one parameter that is critical in the computation of closure$_\Sigma$, that is, the size of Σ, as we will make it precise later.

5   Covers of Dependencies

We now present the different types of covers that are present and adopted in the three fields, namely RDBM, FCA, and Logic. Since the definition of a cover (Definition 3) is very generic, the covers reviewed here have been defined with respect to specific characteristics and for different purposes. As seen above, one of the parameters of both Closure and Linclosure algorithms is a set of dependencies Σ. Then the nature and the size of Σ affect the performance of both functions. On the other hand, the algorithms that compute those covers make use of a function to compute the closure of a set of attributes, for example both functions Closure and Linclosure can be used. Therefore, the implication problem or its equivalent $\text{closure}_{\Sigma}(X)$ and the computation of a cover are two interrelated problems.

5.1   Four Main Characteristics of Covers

In general terms, a cover is simply a set of dependencies Σ. The definition of a cover is very generic and below we introduce some relevant properties which are useful for characterizing the different covers. In particular, Σ is equivalent to another set of dependencies Σ' modulo the Armstrong axioms when the closures Σ^+ and Σ'^+ are the same.

Definition 5. *A set of dependencies Σ is **left-reduced** if and only if for all $X \to Y \in \Sigma$ there is no $X' \to Y$, where $X' \subset X$, such that changing $X \to Y$ by $X' \to Y$ in Σ gives an equivalent base.*

Alternatively, the definition states that $((\Sigma \backslash \{X \to Y\}) \cup \{X' \to Y\})^+ \neq \Sigma^+$ for all $X \to Y \in \Sigma$. This means that all the left-hand sides of all the dependencies in Σ cannot be further reduced to compute the closure Σ^+.

The process of left-reducing a set of dependencies is also mentioned as the removal of **extraneous attributes** in the left-hand sides of all dependencies. As it can be expected, an attribute x is extraneous in the left-hand side of a dependency $\sigma \in \Sigma$ if removing x from the left-hand side in σ does not change Σ^+. The dual definition can be given. An attribute y is extraneous in the right-hand side of a dependency $\sigma \in \Sigma$ if removing y from the right-hand side in σ does not change Σ^+.

Definition 6. *A set of dependencies Σ is **non redundant** if and only if $(\Sigma \backslash \sigma)^+ \neq \Sigma^+$ for all $\sigma \in \Sigma$.*

If a set of dependencies Σ is non redundant and all the right-hand sides are singletons then Σ is **right-reduced**. In fact, non-redundancy and right-reduction are equivalent definitions, as Example 1 shows.

Example 1. Let $\Sigma = \{a \to bc, b \to c, a \to d\}$ be a set of dependencies. The attribute c in the dependency $a \to bc$ is clearly extraneous, since the base $\Sigma' = \{a \to b, b \to c, a \to d\}$ is equivalent to Σ. The transformation of Σ into Σ' can be viewed from two different perspectives: we remove the attribute c in the dependency $a \to bc$ because it is extraneous (right-reduction) or we take the base $\Sigma = \{a \to b, a \to c, b \to c, a \to d\}$ (in which all the right-hand sides are singletons) from which we remove the dependency $a \to c$ because it is redundant (non-redundancy).

Here we keep the expression "non-redundant" instead of "right-reduced" for the sake of clarity. Since $\{X \rightarrow yz\} \equiv \{X \rightarrow y, X \rightarrow z\}$, there is no difference between considering a cover such that all the right-hand sides are singletons, or just *joining* in one single dependency all those dependencies with the same left-hand side. Actually, when the right-hand sides are singletons, the number of dependencies is "artificially" increased.

Definition 7. *A set of dependencies Σ is* **direct** *if and only if* closure$_\Sigma(X)$ *can be computed with a single pass of Σ, for all $X \subseteq \mathcal{U}$.*

This means that computing the closure of any $X \subseteq \mathcal{U}$ implies only one complete exploration of Σ. Actually, this unique exploration represents a very important property, especially when Σ is large or very large. Moreover, it should also be noticed that (i) a cover Σ is direct regardless of how it is represented or sorted, and also (ii) Σ is direct for any attribute subsets $X \subseteq \mathcal{U}$.

Definition 8. *A set of dependencies Σ is* **minimal** *if and only if $|\Sigma| \leq |\Sigma'|$ for all Σ' such that $\Sigma^+ = \Sigma'^+$.*

It is important to notice that when a cover is minimal, it is also non redundant, but the opposite does not necessarily hold. Moreover, let us recall also the importance of computing a non redundant cover given a set of dependencies Σ, and that computing a non redundant cover is equivalent to the logical implication problem.

In the next sections we review three different and major types of covers that are of interest in the RDBM, in Logic, and in FCA.

5.2   The Minimal Cover in the RDBM and Its Variations

We start with the **Minimal Cover**, which is very popular among the RDBM community and can be found in most of the database textbooks under different names (see Table 2).

Table 2. References to the Minimal Cover in RDBM textbooks.

Name	Ref	Where
Canonical Cover	Maier [20]	p. 79, Sect. 5.6
Minimal Cover	Ullman [27]	p. 390
Minimal Cover	Abiteboul [1]	p. 286, Exercice 11.16
Irreducible Set of Dependencies	Date [10]	p. 341, Sect. 11.6
Minimal Cover	Elmasri [12]	p. 549, Sect. 16.1.3
Canonical Cover	Silberschatz [26]	p. 324, Sect. 7.4.3

The computation of the Minimal Cover is performed in three steps:

1. All the dependencies in Σ must have only one single attribute in the right-hand side. This is performed by simply replacing a dependency $X \to Y$ by the dependencies $X \to y_i$ for all $y_i \in Y$.
2. Σ is left-reduced. This is performed by changing a dependency $X \to y$ by a dependency $X' \to y$, where $X' \subset X$, whenever $(\Sigma \setminus \{X \to Y\} \cup \{X' \to Y\})^+ \equiv \Sigma^+$ (see Definition 5).
3. Redundant dependencies are removed from Σ (see Definition 6).

It is important to notice that the order of steps 2 and 3 is relevant and mandatory. Section 5.3 in [20] includes a discussion explaining why left-reduction needs to be performed before the removal of redundant dependencies. The output of computing the Minimal Cover depends also on the order in which dependencies are processed. As a consequence it comes that there may be different minimal covers for the same Σ. Finally, the Minimal Cover does not ensure directness.

Example 2 (Adapted from Sect. 5.2 in [20]).

Let us suppose that we have the following set of dependencies: $\Sigma = \{a \to b, b \to ac, a \to c, bc \to a\}$. Applying step 1 outputs $\Sigma = \{a \to b, b \to a, b \to c, a \to c, bc \to a\}$. Then step 2 is applied to left-reduce Σ. Since $bc \to a$ can be left-reduced to $b \to a$ (thanks to Augmentation), then the following equivalent set is produced: $\Sigma' = \{a \to b, b \to a, b \to c, a \to c\}$. Finally, applying step 3 outputs: $\Sigma'' = \{a \to bc, b \to a\}$. By contrast, let us assume that the order of Σ is changed as follows: $\Sigma = \{a \to b, a \to c, b \to ac, bc \to a\}$. Applying step 1 yields the same set as above: $\Sigma = \{a \to b, a \to c, b \to a, b \to c, bc \to a\}$. When applying step 2, it comes: $\Sigma' = \{a \to b, a \to c, b \to a, b \to c\}$. And finally applying step 3 outputs the base $\Sigma' = \{a \to b, b \to ac\}$.

5.3   The Canonical-Direct Basis

The Canonical-Direct Basis is defined with different characterizations in [8]. Again the computation of this cover is performed in three steps:

1. All the dependencies in Σ must have one single attribute in the right-hand side, as it was already the case above for the computation of the minimal cover. Again this is performed by simply replacing a dependency $X \to Y$ by the dependencies $X \to y_i$ for all $y_i \in Y$.
2. Σ is closed by "pseudo-transitivity", that is, Augmentation plus transitivity. Then dependencies $X \to y$ such that $y \in X$ are removed.
3. Σ is left-reduced (see Definition 5 and step 2 in the construction of the minimal cover).

It can be noticed that there is no removal of redundant dependencies. The only source of redundancy that is taken into account and removed is the one generated by the application of augmentation, but not of transitivity. The Canonical-Direct Basis is not necessarily minimal, nor it is non-redundant, but it is direct.

Example 3. We continue with Example 2: $\Sigma = \{a \to b, b \to ac, a \to c, bc \to a\}$. Applying step 1 produces $\Sigma = \{a \to b, b \to a, b \to c, a \to c, bc \to a\}$. Here, step 2 consists in closing Σ by pseudo-transitivity, which outputs: $\Sigma' = \{a \to b, b \to a, b \to c, a \to c, bc \to a, a \to a, ac \to a, ab \to a, ab \to b, ac \to c\}$. Then, removing the dependencies that can be deduced by Reflexivity, it comes: $\Sigma'' = \{a \to b, b \to a, b \to c, a \to c, bc \to a\}$. When applying step 2 to left-reduce Σ, and since $bc \to a$ can be left-reduced to $b \to a$ (thanks to Augmentation), the following equivalent set is obtained and constitutes the final result: $\Sigma''' = \{a \to b, b \to a, b \to c, a \to c\}$, Since there is no removal of redundant dependencies, then dependency $a \to c$ appears in this canonical-direct basis. By contrast this was not the case in the minimal basis above where $a \to c$ was removed (see Example 2).

We need to notice that this base is also characterized in Formal Concept Analysis by the so called **minimal generators** (or minimal generating set). A minimal generator is defined in [14] (Sect. 2.3.3) as follows: *A generating set of a closed set A is a subset $S \subseteq A$ such that $A = S''$, and, obviously, a minimal generating set of A is a subset of A minimal with respect to this property.* A similar definition exists in [7]: *A minimal generator B of a closed set F is also called a basis for F , i.e. $\phi(B) = F$ and $\phi(A) \subset \phi(B)$ for every $A \subset B$, or a free subset, i.e. for every $x \in B$, $x \in \phi(B \setminus x)$* (where ϕ is a closure operator). What is also relevant is that, in this same paper, the characterization of the canonical direct basis Σ_{cdb} is defined as follows: $\Sigma_{cdb} = \{ B \to \text{closure}_{\Sigma}(B) \setminus B :\ B \subseteq \mathcal{U}$ is a minimal generator of $\text{closure}_{\Sigma}(B) \}$. That is, the left-hand sides of a Canonical-Direct Basis are the set of minimal generators of the implicit closure operator.

5.4   The Duquenne-Guigues basis

The **Duquenne-Guigues basis** [11,15], also called the *Canonical Basis* in the FCA community, is the cover based on pseudo-closed sets [15]. More precisely, tt is defined as follows:

Definition 9. *The **Duquenne-Guigues basis** of a set of dependencies Σ is defined as*
$$\{ X \to \text{closure}_{\Sigma}(X) \mid X \subseteq \mathcal{U} \text{ and } X \text{ pseudoclosed} \}$$

where the definition of a pseudo-closed set of attributes w.r.t. a set of dependencies Σ is:

Definition 10. *Let Σ be a set of dependencies, and $\mathcal{U}$ the related set of attributes. $X \subseteq \mathcal{U}$ is **pseudoclosed** if:*

1. *$X \neq \text{closure}_{\Sigma}(X)$, that is, X is not closed.*
2. *If $Y \subset X$ is a proper subset of X and pseudo-closed, then $\text{closure}_{\Sigma}(Y) \subseteq X$.*

This basis is not direct, but it is minimal and non-redundant. According to [8], this basis is also presented by Maier in [20], where it is called the *Minimum Cover*: "*It has been obtained independently (and with different formulations) by Maier (*Minimum Cover *), and Guigues and Duquenne (*Duquenne-Guigues basis *)*". However, at present, the use and popularity of the Duquenne-Guigues basis seems to be rather restricted within the FCA community [14,15].

Example 4. Let us consider Example 2: $\Sigma = \{\, a \to b, b \to ac, a \to c, bc \to a \,\}$.

As in Σ the pseudo-closed sets are simply $\{\, a \,\}$ and $\{\, b \,\}$, the following Duquenne-Guigues basis is obtained: $\Sigma' = \{\, a \to bc, b \to ac \,\}$.

6 Discussion and Conclusion

For being more complete, we mention without giving more details another basis, namely the **D-Basis** [2], which can be considered as being *between* the Canonical-Direct Basis and the Minimal Cover. This particular basis, developed in lattice theory, while not minimal, is direct. Moreover, it is smaller in size than the canonical-direct basis, that is, D-Basis $\subseteq$ Canonical-Direct Basis.

Relating the three main covers plus the D-Basis is relevant and very interesting. Actually, while the D-Basis and the Canonical-Direct Basis are related by a subset relationship, such a relationship is not known to exist between the Duquenne-Guigues basis and the Minimum Cover, or between the Canonical-Direct Basis and the Duquenne-Guigues basis. In fact, in [8], in the last sentence before the acknowledgements page 28, it is stated that:

> We conclude that this paper is contradicting a conjecture of the literature (in [37]). Indeed, one observes that the premise of implication (10) of Σ_{cd} (a direct cover) is not contained in a premise of any implication of Σ_{can} (the Duquenne-Guigues basis).

This sentence goes back to a conjecture stated in [5] and can be interpreted as follows. The left-hand sides of a Duquenne-Guigues basis, which is minimal, may not be a subset of the left-hand sides of a direct cover.

The RDBM, Logic, and FCA fields are addressing two different, yet related problems: the implication problem and the computation of a compact and representative set –cover or basis– of a complete set of dependencies. Although the first question is solved in the three fields thanks to the same algorithms, that is, Closure or Linclosure, this unanimity disappears in confronting with the choice of a cover. Table 3 summarizes the three types of covers reviewed in Sect. 5, plus the D-Basis. It can be noticed that the Canonical-Direct Basis and the D-Basis do not keep dependencies that can be inferred by the application of the augmentation axiom: $X \to Y \models XZ \to YZ$, while they include dependencies that can be inferred by transitivity. This additional amount of information is enough to make direct these two covers. By contrast, the Minimal Cover does not contain dependencies that can be inferred by augmentation or transitivity as they are

Table 3. Comparing the characteristics of the four bases.

	Canonical Minimal	Canonical Direct	Duquenne-Guigues Minimum	D-Basis
(found in)	RDBM	RDBM, Logic	FCA/RDBM	Lattice Th.
Minimum	no	no	yes	no
Direct	no	yes	no	yes
Redundant	no	yes	no	yes
Unique	no	yes	yes	yes

removed in the last step of the computation. And this explains why the Minimal Cover is not direct.

We have there a kind of "no free lunch theorem": the more information a basis is keeping the more direct the basis can be. By contrast, minimality and non redundancy do not favor directness.

The lack of unanimity can also be noticed in the RDBM only. Indeed textbooks such as [1,10,12,20,26,27] tend to present the Minimal Cover as the preferred cover, while algorithms computing FDs output the Canonical-Direct Basis [24]. This can be interpreted as follows: textbooks are preferring a cover left-reduced and non-redundant, and, hence, containing less and more compact information. However, for discovering FDs, a left-reduced cover is computed without removing redundant dependencies. A possible explanation is that algorithms computing FDs take a dataset as input. Then it is easy to perform a left-reduction w.r.t. the input dataset by removing an attribute from the left-hand side and test whether the dependency still holds. However, a redundancy test is different in the sense that it can only be performed w.r.t. a set of dependencies, *once this set has completely been computed*. Such a test is of a different nature as it is not performed w.r.t. a dataset. Moreover, this does not prevent any algorithm from performing it.

Although the Minimum Cover (or Duquenne-Guigues basis) is also introduced in the RDBM field, it has not enjoyed the same popularity as the Minimal Cover. In fact, the Minimum Cover is introduced and discussed only in Maier [20]. This lack of popularity is probably due to the rather intricate characterization of the Minimum Cover and the related algorithm at that time (see for example Sect. 5.6, Chap. 5 in [20]). This is especially true when compared with the simplicity and expressiveness of the presentation of the Minimal Cover. The characterization of Minimum Cover cannot compete either with the clear characterization of the Duquenne-Guigues basis in FCA, or the simplicity of the NextClosure Algorithm in [14]. However, this "simplicity" is not for free and it comes at the expense of the whole theoretical framework of FCA.

The choice of a cover can be decided depending on the performance that Closure or Linclosure are offering. Both Closure and Linclosure are closely dependent on the "nature" of the set of dependencies Σ. By "nature" we mean the different characteristics explained in Sect. 5, which have an impact on the amount of information as well as on the number of dependencies considered. If Σ is a direct cover (Definition 7), both algorithms have to perform only a single pass of their outer loop. If the cover is not direct, e.g., Canonical-Direct Basis or Duquenne-Guigues basis, then, the number of iterations of the outer loop may be larger, while, at the same time, the number of iterations of the inner loop may be shorter. Again, we are facing the well-known trade-off between "expressivity and complexity": the more expressive in terms of information containment a cover is, the higher the cost of Closure and Linclosure is.

The question of the preference between the Duquenne-Guigues basis and Minimum Cover remains, even if things have changed. What is left for future work is in concern with the practical behavior of the main covers, which has to be compared and investigated from the experimental point of view, especially having in mind the needs in the RDBM, in Logic, and in FCA, but also in knowledge representation and ontology engineering. In particular, attribute exploration [14] may be a good support for evaluating the potential of the main covers studied above in ontology engineering and database design.

Acknowledgements. Jaume Baixeries is funded by a grant AGRUPS-2022 from Universitat Politcnica de Catalunya and is supported by a recognition 2021SGR-Cat (01266 LQMC) from AGAUR (Generalitat de Catalunya). Mehdi Kaytoue and Amedeo Napoli are carrying out this research work as part of the French ANR-21-CE23-0023 Smart-FCA Research Project.

References

1. Abiteboul, S., Hull, R., Vianu, V.: Foundations of Databases. Addison-Wesley, Boston (1995)
2. Adaricheva, K.V., Nation, J.B., Rand, R.: Ordered direct implicational basis of a finite closure system. Discret. Appl. Math. **161**(6), 707–723 (2013)
3. Bazhanov, K., Obiedkov, S.A.: Optimizations in computing the Duquenne-Guigues basis of implications. Ann. Math. Artif. Intell. **70**(1–2), 5–24 (2014)
4. Beeri, C., Bernstein, P.A.: Computational problems related to the design of normal form relational schemas. ACM Trans. Database Syst. **4**(1), 30–59 (1979)
5. Ben-Khalifa, K., Motameny, S.: Horn representation of a concept lattice. Int. J. Gen. Syst. **38**(4), 469–483 (2009)
6. Bernstein, P.A., Swenson, J.R., Tsichritzis, D.C.: A unified approach to functional dependencies and relations. In: Proceedings of ACM SIGMOD, pp. 237–245 (1975)
7. Bertet, K., Demko, C., Viaud, J.F., Guérin, C.: Lattices, closures systems and implication bases: a survey of structural aspects and algorithms. Theor. Comput. Sci. **743**, 93–109 (2016)
8. Bertet, K., Monjardet, B.: The multiple facets of the canonical direct unit implicational basis. Theor. Comput. Sci. **411**(22–24), 2155–2166 (2010)
9. Crama, Y., Hammer, P.L.: Boolean Functions - Theory, Algorithms, and Applications, Encyclopedia of Mathematics and its Applications, vol. 142. Cambridge University Press, Cambridge (2011)
10. Date, C.J.: An Introduction to Database Systems (7. ed.). Addison-Wesley-Longman, Boston (2000)

11. Duquenne, V., Guigues, J.: Mininal family of informative implications resulting from a binary data table. Math. Humanies Sci. **24**, 5–18 (1986)
12. Elmasri, R., Navathe, S.B.: Fundamentals of Database Systems, 3rd Edition. Addison-Wesley-Longman, Boston (2000)
13. Fagin, R., Vardi, M.Y.: The theory of data dependencies — an overview. In: Paredaens, J. (ed.) ICALP 1984. LNCS, vol. 172, pp. 1–22. Springer, Heidelberg (1984). https://doi.org/10.1007/3-540-13345-3_1
14. Ganter, B., Obiedkov, S.A.: Conceptual Exploration. Springer, Berlin (2016). https://doi.org/10.1007/978-3-662-49291-8
15. Ganter, B., Wille, R.: Formal Concept Analysis - Mathematical Foundations. Springer (1999)
16. Horn, A.: On sentences which are true of direct unions of algebras. J. Symb. Log. **16**(1), 14–21 (1951)
17. Ibaraki, T., Kogan, A., Makino, K.: Functional dependencies in horn theories. Artif. Intell. **108**(1–2), 1–30 (1999)
18. Kanellakis, P.C.: Chapter 17 - Elements of Relational Database Theory. In: Van Leeuwen, J. (ed.) Formal Models and Semantics, pp. 1073–1156. Handbook of Theoretical Computer Science, Elsevier, Amsterdam (1990)
19. Kowalski, R.A.: Logic for Problem Solving, The Computer Science Library : Artificial Intelligence Series, vol. 7. North-Holland (1979)
20. Maier, D.: The Theory of Relational Databases. Computer Science Press, New York (1983)
21. Makhalova, T., Buzmakov, A.V., Kuznetsov, S.O., Napoli, A.: Introducing the closure structure and the GDPM algorithm for mining and understanding a tabular dataset. Int. J. Approximate Reasoning **145**, 75–90 (2022)
22. Mannila, H., Räihä, K.: Design of Relational Databases. Addison-Wesley, Boston (1992)
23. Padawitz, P.: Computing in Horn Clause Theories, EATCS Monographs on Theoretical Computer Science, vol. 16. Springer, Berlin (1988). https://doi.org/10.1007/978-3-642-73824-1
24. Papenbrock, T., et al.: Functional dependency discovery: an experimental evaluation of seven algorithms. Proc. VLDB Endow. **8**(10), 1082–1093 (2015)
25. Sagiv, Y., Delobel, C., Jr., D.S.P., Fagin, R.: An equivalence between relational database dependencies and a fragment of propositional logic. J. ACM **28**(3), 435–453 (1981)
26. Silberschatz, A., Korth, H.F., Sudarshan, S.: Database System Concepts, Seventh Edition. McGraw-Hill Book Company, New York (2020)
27. Ullman, J.D.: Principles of Database and Knowledge-Base Systems, Volume I, Principles of computer science series, vol. 14. Computer Science Press, New York (1988)
28. Wild, M.: Computations with finite closure systems and implications. In: Du, D.-Z., Li, M. (eds.) COCOON 1995. LNCS, vol. 959, pp. 111–120. Springer, Heidelberg (1995). https://doi.org/10.1007/BFb0030825

A Triadic Generalisation of the Boolean Concept Lattice

Alexandre Bazin[✉]

LIRMM, CNRS, Université de Montpellier, Montpellier, France
alexandre.bazin@umontpellier.fr

Abstract. Boolean concept lattices are fundamental structures in formal concept analysis, both from a theoretical and an applied point of view. There are multiple ways to generalise them in the triadic concept analysis framework and one of them, the so-called powerset trilattice, has already been proposed by Biedermann in 1998. However, it lacks some interesting properties such as extremality in the number of triconcepts for tricontexts of a given size. In this paper, we discuss another generalisation of Boolean concept lattices that exhibit such properties. We argue that those structures form equivalence classes and should be studied as such, and investigate the minimum number of objects required to produce them.

Keywords: Formal Concept Analysis · Triadic Concept Analysis · Boolean lattice

1 Introduction

Formal Concept Analysis (FCA [9]) is a formalism that establishes a connection between classical binary data (crosstables) and the structure of concepts and rules that can be found in said data. It is very powerful, if underutilised, as it offers well-studied mathematical structures to be exploited by algorithms.

As crosstables are a rather limiting way of representing data, various extensions of the formalism have been proposed to deal with more complex data, such as Pattern Structures [7], Relational Concept Analysis [12], fuzzy FCA [11] or graph FCA [6]. Just like FCA, they are based on lattice theory [5].

Triadic Concept Analysis [10] is another such formalism that aims to extend FCA to data in the form of ternary relations (*i.e.* tridimensional crosstables), and has the peculiarity of involving trilattices instead of lattices. Such structures are much less studied and many questions remain open (or waiting to be opened).

In this paper, we are interested in the tridimensional generalisation of the well known Boolean concept lattices. There are multiple ways to generalise these structures to the tridimensional case, one of them having been studied by Biedermann [3]. While interesting in many aspects, these so-called powerset trilattices lack properties that are of interest in pattern mining. For instance, Boolean concept lattices contain all possible subsets of attributes as intents, *i.e.* all possible descriptions, making it the fundamental search space for anything itemset-related. This is not the case of powerset trilattices

D. Dürrschnabel and D. López Rodríguez (Eds.): ICFCA 2023, LNAI 13934, pp. 95–105, 2023.
https://doi.org/10.1007/978-3-031-35949-1_7

as its triconcepts do not contain all possible combinations of attributes and conditions. Powerset trilattices are also not extremal in the number of triconcepts for tricontexts of a given size.

We thus discuss here another generalisation of Boolean concept lattices that exhibits the properties we are looking for. In Sect. 3, we begin by arguing that the absence of duality between the quasi-orders of a trilattice opens a new way to study these structures: as classes of trilattices sharing the same structure of "descriptions" of triconcepts. In Sect. 4, we define the classes of trilattices generalising Boolean concept lattices, and in Sect. 5 we discuss the minimum number of objects required to produce such trilattices.

2   Formal and Triadic Concept Analysis

2.1   Formal Concept Analysis

We begin by presenting the essential notions of formal concept analysis. For an indepth introduction to FCA, we refer the interested reader to the book [9]. A formal context is a triple $(\mathcal{O}, \mathcal{A}, \mathcal{I})$ where $\mathcal{O}$ is a set of *objects*, $\mathcal{A}$ is a set of *attributes* and $\mathcal{I} \subseteq \mathcal{O} \times \mathcal{A}$ is a binary relation between objects and attributes. We say that object o is *described* by attribute a when $(o, a) \in \mathcal{I}$. The *description* of the object o is thus a set of attributes. Formal contexts can be represented as crosstables, as depicted in Fig. 1. A formal context is said to be *reduced* when no row (resp. column) is empty, full of crosses or equal to the intersection of other rows (resp. columns). We suppose in this paper that all contexts are reduced.

	a_1	a_2	a_3	a_4	a_5
o_1	×	×			
o_2		×	×	×	
o_3		×		×	×
o_4	×		×		
o_5				×	×

Fig. 1. A crosstable representing a formal context with five objects ($\{o_1, o_2, o_3, o_4, o_5\}$) and five attributes ($\{a_1, a_2, a_3, a_4, a_5\}$).

From a formal context, two derivation operators $\cdot'$ are defined such that

- $\cdot' : 2^{\mathcal{O}} \to 2^{\mathcal{A}}, O \mapsto O' = \{a \in \mathcal{A} \mid \forall o \in O, (o, a) \in \mathcal{I}\}$, and
- $\cdot' : 2^{\mathcal{A}} \to 2^{\mathcal{O}}, A \mapsto A' = \{o \in \mathcal{O} \mid \forall a \in A, (o, a) \in \mathcal{I}\}$.

A *formal concept* is a pair $(E, I) \in 2^{\mathcal{O}} \times 2^{\mathcal{A}}$ such that $E = I'$ and $I = E'$. We call E the *extent* of the concept while I is its *intent*. Concepts correspond to maximal rectangles of crosses in the formal context, up to permutation of rows and columns. Hence, another definition is that a formal concept is a pair (E, I) such that $E \times I \in \mathcal{I}$ and both components are maximal for this property. According to the basic theorem of formal concept analysis, the set of formal concepts in a formal context ordered by the inclusion relation on either their intents or extents forms a complete lattice called the *concept lattice* of the context (Fig. 2).

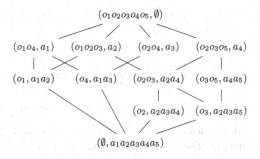

Fig. 2. Concept lattice of the formal context depicted in Fig. 1.

The Boolean concept lattices are the concept lattices of contexts of the form $(S, S, \neq)$ called contranominal scales. Figure 3 depicts such a Boolean concept lattice for $|S| = 3$. In a Boolean concept lattice, all subsets of attributes (resp. objects) are intents (resp. extents) of concepts. As such, it contains $2^{|S|}$ concepts.

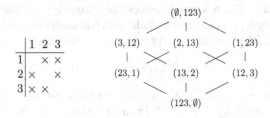

Fig. 3. A Boolean concept lattice

2.2 Triadic Concept Analysis

Triadic concept analysis is the tridimensional generalisation of FCA introduced by Lehman and Wille [10]. It has since then been further generalised to the n-dimensional case by Voutsadakis [13] but this is outside the scope of this paper. In this setting, a triadic context (or tricontext, or 3-context) is a triple $(\mathcal{O}, \mathcal{A}, \mathcal{C}, \mathcal{I})$ where the $\mathcal{O}$ is a set of object, $\mathcal{A}$ is a set of attributes, $\mathcal{C}$ is a set of conditions and $\mathcal{I} \subseteq \mathcal{O} \times \mathcal{A} \times \mathcal{C}$ is a ternary relation between elements of the three dimensions. In this paper, we shall say that the pairs (a, c) such that $(o, a, c) \in \mathcal{I}$ form the *description* of the object o. An example of a triadic context is depicted in Fig. 4. For practical reasons that will be clarified later, we represent the object dimension on the bottom of the figure so that the descriptions of objects are visually represented as classical formal contexts.

A triadic concept (or triconcept, or 3-concept) is then a maximal tridimensional box full of crosses, *i.e.* a triple (X_1, X_2, X_3) such that $X_1 \times X_2 \times X_3 \subseteq \mathcal{I}$ and the X_i are maximal for this property. For instance, $(\{o_1, o_2\}, \{a_1, a_2\}, \{c_1\})$ is a triconcept in

	c_1	c_2	c_3		c_1	c_2	c_3		c_1	c_2	c_3
a_1	×				×		×				×
a_2	×				×						
a_3	×	×				×				×	×
		o_1				o_2				o_3	

Fig. 4. A triadic context $(\{o_1, o_2, o_3\}, \{a_1, a_2, a_3\}, \{c_1, c_2, c_3\}, \mathcal{I})$.

Fig. 4. We shall say that, in a triconcept (O, A, C), O is the extent while (A, C) is the description of the triconcept. The set $\mathcal{S}$ of all triconcepts in a tricontext together with the three quasi-orders

$$(X_1, X_2, X_3) \lesssim_i (Y_1, Y_2, Y_3) \Leftrightarrow X_i \subseteq Y_i, \ i \in \{1, 2, 3\},$$

forms a triadic lattice (or trilattice, or 3-lattice) $\mathcal{L} = (\mathcal{S}, \lesssim_1, \lesssim_2, \lesssim_3)$ [4] called the concept trilattice of the tricontext. A trilattice is a triordered set, *i.e.*:

– if $\forall i \in \{1, 2, 3\} \backslash \{j\}, A \lesssim_i B$ then $A \gtrsim_j B$ (antiordinal dependency) and
– if $\forall i \in \{1, 2, 3\}, A \sim_i B$, then $A = B$ (uniqueness condition).

The duality of the two partial orders on extents and intents in the dyadic case is thus lost in the triadic case. This has consequences that are discussed in the next section.

In [3], Biedermann proposed a first triadic generalisation of Boolean lattices called *powerset trilattices*. They are the concept trilattices of tricontexts of the form

$$(S, S, S, S^3 \backslash \{(a, a, a) \mid a \in S\}).$$

They contain $3^{|S|}$ triconcepts (X_1, X_2, X_3) that are such that $X_1 \cap X_2 \cap X_3 = \emptyset$ and $X_i \cup X_j = S$ for distinct $i, j \in \{1, 2, 3\}$. Figure 5 depicts such a powerset trilattice for $|S| = 3$.

	1	2	3		1	2	3		1	2	3
1		×	×		×	×	×		×	×	×
2	×	×	×		×		×		×	×	×
3	×	×	×		×	×	×		×	×	
		1				2				3	

$(\{1\}, \{1, 2, 3\}, \{2, 3\})$	$(\{1\}, \{2, 3\}, \{1, 2, 3\})$	$(\{2\}, \{1, 2, 3\}, \{1, 3\})$
$(\{2\}, \{1, 3\}, \{1, 2, 3\})$	$(\{3\}, \{1, 2\}, \{1, 2, 3\})$	$(\{3\}, \{1, 2, 3\}, \{1, 2\})$
$(\{1, 3\}, \{1, 2\}, \{2, 3\})$	$(\{1, 3\}, \{2, 3\}, \{1, 2\})$	$(\{2, 3\}, \{1, 2\}, \{1, 3\})$
$(\{1, 2\}, \{2, 3\}, \{1, 3\})$	$(\{2, 3\}, \{1, 3\}, \{1, 2\})$	$(\{1, 2\}, \{1, 3\}, \{2, 3\})$
$(\{2, 3\}, \{1, 2, 3\}, \{1\})$	$(\{1, 2\}, \{1, 2, 3\}, \{2\})$	$(\{1, 2\}, \{1, 2, 3\}, \{3\})$
$(\{2, 3\}, \{1\}, \{1, 2, 3\})$	$(\{1, 3\}, \{2\}, \{1, 2, 3\})$	$(\{1, 2\}, \{3\}, \{1, 2, 3\})$
$(\{1, 2, 3\}, \{1\}, \{2, 3\})$	$(\{1, 2, 3\}, \{2\}, \{1, 3\})$	$(\{1, 2, 3\}, \{3\}, \{1, 2\})$
$(\{1, 2, 3\}, \{2, 3\}, \{1\})$	$(\{1, 2, 3\}, \{1, 3\}, \{2\})$	$(\{1, 2, 3\}, \{1, 2\}, \{3\})$
$(\emptyset, \{1, 2, 3\}, \{1, 2, 3\})$	$(\{1, 2, 3\}, \emptyset, \{1, 2, 3\})$	$(\{1, 2, 3\}, \{1, 2, 3\}, \emptyset)$

Fig. 5. The tricontext and triconcepts of a powerset trilattice on three elements.

3 The Loss of Duality in Triadic Concept Analysis

The biggest difference between the bidimensional and the tridimensional cases is that, in a concept lattice, the partial order on extents is dual to the partial order on intents whereas fixing two quasi-orders in a concept trilattice does not determine the third quasi-order. This is exemplified in Fig. 6 where two different tricontexts produce two sets of triconcepts that have the same descriptions but different extents. The second and third quasi-orders of both trilattices are thus isomorphic while the two first quasi-orders are not.

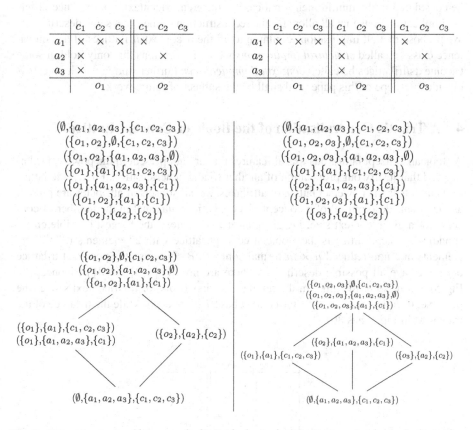

Fig. 6. Two tricontexts $(\{o_1, o_2, o_3\}, \{1, 2, 3\}, \{a, b, c\}, \mathcal{I}_1)$ (left) and $(\{o_1, o_2\}, \{1, 2, 3\}, \{a, b, c\}, \mathcal{I}_2)$ (right), their associated triconcepts and the two first quasi-orders.

For a trilattice $\mathcal{L} = (\mathcal{S}, \lesssim_1^{\mathcal{L}}, \lesssim_2^{\mathcal{L}}, \lesssim_3^{\mathcal{L}})$, we call $\mathcal{L}^{\equiv}$ the equivalence class comprised of all trilattices $\mathcal{L}_2 = (\mathcal{S}_2, \lesssim_1^{\mathcal{L}_2}, \lesssim_2^{\mathcal{L}_2}, \lesssim_3^{\mathcal{L}_2})$ such that $\lesssim_i^{\mathcal{L}}$ is isomorphic to $\lesssim_i^{\mathcal{L}_2}$ for $i \in \{2, 3\}$. We argue that the study of such equivalence classes in triadic and polyadic concept analysis opens up new and interesting questions, even beyond the purely mathematical aspects:

- what does it mean from the point of view of knowledge representation when there are different ways objects can belong to formal concepts as defined by their descriptions?
- if the size of a dimension can be changed without modifying the structure, can it lead to new data augmentation/reduction techniques?
- if a tricontext produces a trilattice with the minimum number of objects among its equivalence class, does it mean anything?

Brief Divagation. These equivalence classes impact the way implications [2, 8] should be considered in the multidimensional case as different tricontexts can produce different implications sets that all allow for the reconstruction of the exact same descriptions. We propose that all implications common to all the tricontexts/trilattices of an equivalence class be called *structural implications* while implications that only hold in some tricontexts/trilattices be called *contextual implications*. Further study of such objects is outside the scope of this paper and shall be the subject of future work.

4   A Triadic Generalisation of the Boolean Concept Lattice

A Boolean concept lattice has useful features that are not present in the powerset trilattice and that justify our proposition of another triadic generalisation. Firstly, the lattice of its intents is the powerset lattice of attributes, *i.e.* all subsets of attributes are present as the intent (description) of a concept. Secondly, it is maximal in the number of concepts for a given context size, *i.e.* given an $n \times n$ context, the biggest possible corresponding concept lattice is the Boolean concept lattice with 2^n elements ($2^{min(n,m)}$ elements in a non-reduced $n \times m$ formal context). By contrast, the powerset trilattice does not have all possible descriptions (there are no $(X, \{2, 3\}, \{2, 3\})$ triconcept in Fig. 5) and it is not maximal in the number of triconcepts for its tricontext's size (the powerset trilattice on a $3 \times 3 \times 3$ tricontext has 27 triconcepts while the trilattice of the tricontext in Fig. 7 has 30).

	c_1 c_2 c_3	c_1 c_2 c_3	c_1 c_2 c_3
a_1	× ×	× ×	× ×
a_2	× ×	× ×	× ×
a_3	× ×	× ×	× ×
	o_1	o_2	o_3

Fig. 7. A $3 \times 3 \times 3$ tricontext that contains more triconcepts than the $3 \times 3 \times 3$ powerset trilattice.

Having all possible descriptions in a concept lattice is useful as a basic search space for pattern mining. Most itemset mining algorithms can be seen as cleverly searching the powerset of items, so having the multidimensional equivalent of this search space would allow for similar new multidimensional description mining algorithms. As for the maximal number of concepts in fixed contexts, its knowledge is required in the rigorous analysis of algorithms. We also expect these structures to play a role in the study of "big" trilattices, similarly to Boolean concept lattices [1]. Therefore, we propose the following triadic generalisations of the Boolean concept lattice.

Let $\mathcal{A}$ and $\mathcal{C}$ be an attribute and a condition set and $\mathcal{L} = (S, \lesssim_1, \lesssim_2, \lesssim_3)$ be a concept trilattice such that

$$(X_1, X_2, X_3) \in S \Leftrightarrow (X_2, X_3) \in ((2^{\mathcal{A}} \backslash \emptyset) \times (2^{\mathcal{C}} \backslash \emptyset)) \cup \{(\emptyset, \mathcal{C}), (\mathcal{A}, \emptyset)\}.$$

This trilattice is a generalisation of the Boolean concept lattice as it contains all possible concept description and, as two components of a concept uniquely define the third, it is thus maximal in the number of concepts among all trilattices on tricontexts of the form $(\mathcal{O}, \mathcal{A}, \mathcal{C}, \mathcal{I})$. However, as discussed in the previous section, there are multiple such trilattices for a given $\mathcal{A}$ and $\mathcal{C}$. Hence, for $n = |\mathcal{A}|$ and $m = |\mathcal{C}|$, we use $\mathcal{B}_{n,m} = \mathcal{L}^{\equiv}$ to denote the equivalence class of trilattices that contain all possible descriptions, *i.e.* all possible rectangles, in an $n \times m$ crosstable. More formally, the concept trilattices $(S, \lesssim_1, \lesssim_2, \lesssim_3)$ in $\mathcal{B}_{n,m}$ are the concept trilattices of tricontexts $(\mathcal{O}, \mathcal{A}, \mathcal{C}, \mathcal{I})$ (we still suppose that all contexts are reduced) such that $|\mathcal{A}| = n$, $|\mathcal{C}| = m$ and, for all $(Y, Z) \in (2^{\mathcal{A}} \backslash \emptyset) \times (2^{\mathcal{C}} \backslash \emptyset)$, there is a triconcept $(X, Y, Z) \in S$. Clearly, the powerset trilattice depicted in Fig. 5 does not belong to $\mathcal{B}_{3,3}$ as it has no triconcept $(X, \{3\}, \{2, 3\})$.

Figure 8 depicts the tricontexts and triconcepts of two elements of $\mathcal{B}_{2,2}$. We can see that the 9 non-empty rectangles in a 2×2 crosstable, as well as two occurrences of empty rectangles, are present as descriptions of triconcepts.

	$c_1\ c_2$	$c_1\ c_2$	$c_1\ c_2$	$c_1\ c_2$
a_1	× ×		×	×
a_2		× ×	×	×
	o_1	o_2	o_3	o_4

	$c_1\ c_2$	$c_1\ c_2$	$c_1\ c_2$
a_1	× ×	×	×
a_2		×	× ×	×
	o_1	o_2	o_3

$(\emptyset, \{a_1, a_2\}, \{c_1, c_2\})$
$(\{o_1\}, \{a_1\}, \{c_1, c_2\})$
$(\{o_2\}, \{a_2\}, \{c_1, c_2\})$
$(\{o_3\}, \{a_1, a_2\}, \{c_1\})$
$(\{o_4\}, \{a_1, a_2\}, \{c_2\})$
$(\{o_1, o_3\}, \{a_1\}, \{c_1\})$
$(\{o_1, o_4\}, \{a_1\}, \{c_2\})$
$(\{o_2, o_3\}, \{a_2\}, \{c_1\})$
$(\{o_2, o_4\}, \{a_2\}, \{c_2\})$
$(\{o_1, o_2, o_3, o_4\}, \emptyset, \{c_1, c_2\})$
$(\{o_1, o_2, o_3, o_4\}, \{a_1, a_2\}, \emptyset)$

$(\emptyset, \{a_1, a_2\}, \{c_1, c_2\})$
$(\{o_1\}, \{a_1\}, \{c_1, c_2\})$
$(\{o_2\}, \{a_2\}, \{c_1, c_2\})$
$(\{o_2\}, \{a_1, a_2\}, \{c_1\})$
$(\{o_1\}, \{a_1, a_2\}, \{c_2\})$
$(\{o_1, o_2\}, \{a_1\}, \{c_1\})$
$(\{o_1, o_3\}, \{a_1\}, \{c_2\})$
$(\{o_2, o_3\}, \{a_2\}, \{c_1\})$
$(\{o_1, o_2\}, \{a_2\}, \{c_2\})$
$(\{o_1, o_2, o_3\}, \emptyset, \{c_1, c_2\})$
$(\{o_1, o_2, o_3\}, \{a_1, a_2\}, \emptyset)$

Fig. 8. Two tricontexts producing trilattices in $\mathcal{B}_{2,2}$.

A trilattice in $\mathcal{B}_{n,m}$ contains $(2^n - 1) \times (2^m - 1) + 2$ triconcepts, corresponding to the $(2^n - 1) \times (2^m - 1)$ descriptions that are non-empty rectangles plus two times the empty rectangle in the two triconcepts of the form $(\mathcal{O}, \emptyset, \mathcal{C})$ and $(\mathcal{O}, \mathcal{A}, \emptyset)$.

5 On the Minimum Number of Objects in a $\mathcal{B}_{n,m}$ Tricontext

In Fig. 8, we see two $\mathcal{B}_{2,2}$ trilattices, one built on 3 objects and the other on 4. A question then naturally arises: what is the minimum number of objects required to produce a $\mathcal{B}_{n,m}$ trilattice? In the bidimensional case, Boolean concept lattices are produced by contranominal scales and require as many objects as attributes. In the triadic case, it is easy to see that any $\mathcal{B}_{n,m}$ trilattice can be produced on $n+m$ objects. Indeed, the rectangles missing only a single row or column are *irreducible*, *i.e.* they cannot be obtained by intersecting two other rectangles, so they have to appear as maximal rectangles in the description of an object. Having all these rectangles on different objects produces a $\mathcal{B}_{n,m}$ trilattice on $n+m$ objects, as exemplified on the left hand side of Fig. 8 and in Fig. 9. Can we do it with fewer objects? In the case of $\mathcal{B}_{2,2}$ yes, as shown in Fig. 8 with 3 objects, but the answer is not so simple for other sizes.

First of all, we know that if (X_1, X_2, X_3) and (Y_1, Y_2, Y_3) are triconcepts such that $X_2 \subseteq Y_2$ and $X_3 \subseteq Y_3$, then $Y_1 \subseteq X_1$. As the height of the poset of rectangles in an $n \times m$ crosstable ordered by inclusion is $n+m$, we have that a $\mathcal{B}_{n,m}$ trilattice cannot be produced on fewer than $n+m-1$ objects. The minimum number of objects thus lies in $[n+m-1, n+m]$. We now inelegantly show that it is $n+m$ when either $n > 2$ or $m > 2$.

	c_1 c_2 c_3	c_1 c_2 c_3	c_1 c_2 c_3	c_1 c_2 c_3	c_1 c_2 c_3	c_1 c_2 c_3
a_1	× × ×	× × ×		× ×	× ×	× ×
a_2	× × ×		× × ×	× ×	× ×	× ×
a_3		× × ×	× × ×	× ×	× ×	× ×
	o_1	o_2	o_3	o_4	o_5	o_6

Fig. 9. A $6 \times 3 \times 3$ $\mathcal{B}_{3,3}$ tricontext.

Let us start with the case $n = m$ and illustrate with $n = m = 3$ without loss of generality. As the $n+m$ irreducible rectangles have to appear somewhere, if we want to construct a $\mathcal{B}_{n,m}$ tricontext on $n+m-1$ objects, we have to put two irreducible rectangles on the same object. For instance, $(\{a_1, a_2\}, \{c_1, c_2, c_3\})$ and $(\{a_1, a_2, a_3\}, \{c_2, c_3\})$.

	c_1 c_2 c_3	c_1 c_2 c_3	c_1 c_2 c_3	c_1 c_2 c_3	c_1 c_2 c_3	c_1 c_2 c_3
a_1	× × ×	× × ×		× ×	× ×	
a_2	× × ×		× × ×	× ×	× ×	
a_3	× ×	× × ×	× × ×	× ×	× ×	
	o_1	o_2	o_3	o_4	o_5	o_6

However, if we do this, their intersection $(\{a_1, a_2\}, \{c_2, c_3\})$ is not the description of a triconcept anymore. The $(\{a_1, a_2\}, \{c_2, c_3\})$ rectangle must thus appear as a maximal rectangle in the description of another object. The only option is to put it on its own object and then we go back to having $n+m$ objects. From there, being able to construct a $\mathcal{B}_{n,m}$ tricontext on $n+m-1$ objects implies being able to construct a $\mathcal{B}_{n-1,m-1}$ tricontext on $n+m-3$ objects (in the greyed area) as two objects are already taken by the two irreducible rectangles and their intersection.

	c_1	c_2	c_3	c_1	c_2	c_3	c_1	c_2	c_3	c_1	c_2	c_3	c_1	c_2	c_3
a_1	X	X	X	X	X										
a_2	X	X	X	X	X										
a_3		X	X												
	o_1			o_2			o_3			o_4			o_5		

It is possible to construct a $\mathcal{B}_{2,2}$ tricontext with 3 objects (see Fig. 8): two pairs of irreducible rectangles are put on two objects and both their intersections coexist on a third object. To show that this is only one way to do so, up to permutation of rows and columns, is left as an exercise for the reader. The construction of a hypothetical $\mathcal{B}_{3,3}$ tricontext on 5 objects would thus proceed as follows:

	c_1	c_2	c_3	c_1	c_2	c_3	c_1	c_2	c_3	c_1	c_2	c_3	c_1	c_2	c_3
a_1	X	X	X	X	X		X	X		X					X
a_2	X	X	X	X	X				X	X	X		X		
a_3		X	X												
	o_1			o_2			o_3			o_4			o_5		

As the irreducible rectangles still need to appear somewhere, the only way to have them appear without destroying the $\mathcal{B}_{2,2}$ subcontext is thus as follows:

	c_1	c_2	c_3	c_1	c_2	c_3	c_1	c_2	c_3	c_1	c_2	c_3	c_1	c_2	c_3
a_1	X	X	X	X	X		X	X	X	X	X				X
a_2	X	X	X	X	X		X		X	X	X	X	X		
a_3		X	X				X	X	X	X	X	X			
	o_1			o_2			o_3			o_4			o_5		

The coexistence of irreducible rectangles on the same object again causes their intersection to stop being descriptions of triconcepts, so we have to add $(\{a_1,a_3\},\{c_1,c_3\})$ and $(\{a_2,a_3\},\{c_1,c_2\})$ as maximal rectangles in the description of an object. One of them can be added to the description of o_5 but it can be seen that none of them can be added to the description of o_2. For example, adding $(\{a_2,a_3\},\{c_1,c_2\})$ would cause $(\{a_2\},\{c_2,c_3\})$ to stop being the description of a triconcept, creating the need for a sixth object. We are thus stuck, and conclude that $\mathcal{B}_{3,3}$ trilattices require at least 6 objects. From this we deduce that $\mathcal{B}_{n,n}$ trilattices require at least $2n$ objects (Fig. 10).

	c_1	c_2	c_3	c_1	c_2	c_3	c_1	c_2	c_3	c_1	c_2	c_3	c_1	c_2	c_3	c_1	c_2	c_3
a_1	X	X	X	X	X		X	X	X	X	X		X		X			
a_2	X	X	X	X	X		X		X	X	X	X	X			X	X	
a_3		X	X				X	X	X	X	X	X	X		X	X	X	
	o_1			o_2			o_3			o_4			o_5			o_6		

Fig. 10. Another way to build a $\mathcal{B}_{3,3}$ trilattice on 6 objects.

Now, we show analogously that $\mathcal{B}_{2,m}$ trilattices require at least $2 + m$ objects. We illustrate on a $\mathcal{B}_{2,3}$ trilattice but assert that the same construction and reasoning apply to $m > 3$. As for the 3×3 case, all irreducible rectangles in a $2 \times m$ crosstable have to appear in the description of an object.

	c_1	c_2	c_3		c_1	c_2	c_3		c_1	c_2	c_3		c_1	c_2	c_3		c_1	c_2	c_3
a_1	×	×			×		×			×	×		×	×	×				
a_2	×	×			×		×			×	×						×	×	×
		o_1				o_2				o_3				o_4				o_5	

In order to reduce the number of objects, two such rectangles have to be put on the same object. Their intersection ceases to be the description of a triconcept so we have to put it on its own object.

	c_1	c_2	c_3		c_1	c_2	c_3		c_1	c_2	c_3		c_1	c_2	c_3
a_1	×	×	×		×	×									
a_2	×	×													
		o_1				o_2				o_3				o_4	

If it is possible to construct a $\mathcal{B}_{2,m}$ trilattice on $2 + m - 1$ objects, then it is possible to construct a $\mathcal{B}_{1,m-1}$ trilattice on $2 + m - 3$ objects. It is the case here. In order to complete the tricontext so that all irreducible rectangles appear as maximal rectangles in the description of an object, the unused horizontal irreducible rectangle $((\{a_2\}, \{c_1, c_2, c_3\})$ here) has to coexist with another irreducible rectangle. Their intersection thus has to appear as a maximal rectangle in the description of another object and it easy to see, on such a small example, that it is impossible without having rectangles cease to be descriptions of triconcepts (adding it to o_2 would cause $(\{a_1\}, \{c_1\})$ to stop being the description of a triconcept).

	c_1	c_2	c_3		c_1	c_2	c_3		c_1	c_2	c_3		c_1	c_2	c_3		c_1	c_2	c_3
a_1	×	×	×		×	×			×		×		×	×					
a_2	×	×							×	×	×		×	×		×			×
		o_1				o_2				o_3				o_4				o_5	

Hence, $\mathcal{B}_{n,m}$ trilattices require at least $n + m$ objects when either $n > 2$ or $m > 2$.

6 Discussion and Conclusion

The generalisation of Boolean concept lattices proposed in this paper is only one of many but we feel that its properties are interesting for both pattern mining and the study of the maximal size of concept trilattices. The latter is still an open question as, for instance, powerset trilattices are not the biggest trilattices that can be built in a $n \times n \times n$ tricontexts and do not fit. The next step would be to further generalise these structures to the n-dimensional case. The definition is straightforward but our proof of the minimum number of objects required to build the trilattice is hardly scalable as it relies on handmade contexts that would be too big and numerous in higher dimensions. We also plan on further studying the structure of the classes of trilattices/n-lattices defined in Sect. 3 and the consequences of that loss of duality on implications.

Acknowledgements. This work was supported by the ANR SmartFCA project Grant ANR-21-CE23-0023 of the French National Research Agency.

References

1. Albano, A., Chornomaz, B.: Why concept lattices are large: extremal theory for generators, concepts, and VC-dimension. Int. J. Gen Syst **46**(5), 440–457 (2017)
2. Bazin, A.: On implication bases in n-lattices. Discrete Appl. Math. **273**, 21–29 (2020)
3. Biedermann, K.: Powerset trilattices. In: Mugnier, M.-L., Chein, M. (eds.) ICCS-ConceptStruct 1998. LNCS, vol. 1453, pp. 209–221. Springer, Heidelberg (1998). https://doi.org/10.1007/BFb0054916
4. Biedermann, K.: An equational theory for trilattices. Algebra Univers. **42**, 253–268 (1999). https://doi.org/10.1007/s000120050002
5. Birkhoff, G.: Lattice Theory, vol. 25. American Mathematical Society (1940)
6. Ferré, S., Cellier, P.: Graph-FCA: an extension of formal concept analysis to knowledge graphs. Discrete Appl. Math. **273**, 81–102 (2020)
7. Ganter, B., Kuznetsov, S.O.: Pattern structures and their projections. In: Delugach, H.S., Stumme, G. (eds.) ICCS-ConceptStruct 2001. LNCS (LNAI), vol. 2120, pp. 129–142. Springer, Heidelberg (2001). https://doi.org/10.1007/3-540-44583-8_10
8. Ganter, B., Obiedkov, S.: Implications in triadic formal contexts. In: Wolff, K.E., Pfeiffer, H.D., Delugach, H.S. (eds.) ICCS-ConceptStruct 2004. LNCS (LNAI), vol. 3127, pp. 186–195. Springer, Heidelberg (2004). https://doi.org/10.1007/978-3-540-27769-9_12
9. Ganter, B., Wille, R.: Formal Concept Analysis: Mathematical Foundations. Springer, Heidelberg (2012)
10. Lehmann, F., Wille, R.: A triadic approach to formal concept analysis. In: Ellis, G., Levinson, R., Rich, W., Sowa, J.F. (eds.) ICCS-ConceptStruct 1995. LNCS, vol. 954, pp. 32–43. Springer, Heidelberg (1995). https://doi.org/10.1007/3-540-60161-9_27
11. Poelmans, J., Ignatov, D.I., Kuznetsov, S.O., Dedene, G.: Fuzzy and rough formal concept analysis: a survey. Int. J. Gen Syst **43**(2), 105–134 (2014)
12. Rouane-Hacene, M., Huchard, M., Napoli, A., Valtchev, P.: Relational concept analysis: mining concept lattices from multi-relational data. Ann. Math. Artif. Intell. **67**(1), 81–108 (2013). https://doi.org/10.1007/s10472-012-9329-3
13. Voutsadakis, G.: Polyadic concept analysis. Order **19**(3), 295–304 (2002)

Applications and Visualization

Computing Witnesses for Centralising Monoids on a Three-Element Set

Mike Behrisch[1,2(✉)] [iD] and Leon Renkin[1,3] [iD]

[1] Institute of Discrete Mathematics and Geometry, Technische Universität Wien, Vienna, Austria
[2] Institut für Algebra, Johannes Kepler Universität Linz, Linz, Austria
behrisch@logic.at, mike.behrisch@jku.at
[3] Department of Mathematics, ETH Zürich, Zürich, Switzerland

Abstract. We use a formal concept theoretic approach to computationally validate the classification of all 192 centralising monoids on three-element sets by Machida and Rosenberg. We determine a manageable finite (and row-reduced) context the intents of which are exactly all centralising monoids on $\{0, 1, 2\}$. As an advantage of our method, we are able to compute a list with a witness for each of the 192 monoids, which has not been available to date and allows for an immediate verification of the figure 192 as a lower bound for the number of centralising monoids. We also confirm that the 192 centralising monoids split into 48 conjugacy classes and we provide a list of all conjugacy types including witnesses.

Keywords: Centralising monoid · Witness · Context reduction

1 Introduction

Commutation of unary functions $f, g\colon A \to A$ in the sense $f(g(x)) = g(f(x))$ for all $x \in A$ has a natural generalisation to operations $f\colon A^n \to A$ and $g\colon A^m \to A$ of higher arities $m, n \in \mathbb{N}$. We say that f and g *commute*, and denote this as $f \perp g$, if for every matrix $X \in A^{m \times n}$ with rows $r_1, \ldots, r_m \in A^n$ and columns $c_1, \ldots, c_n \in A^m$ the following equation with $m \cdot n$ arguments holds:

$$f(g(c_1), \ldots, g(c_n)) = g(f(r_1), \ldots, f(r_m)). \tag{†}$$

If we write the row-wise application $(f(r_1), \ldots, f(r_m))$ of f to the matrix X as $f(X)$ and if $X^\top$ denotes the transpose of X, then we may restate the commutation property in the form $f(g(X^\top)) = g(f(X))$ for all $X \in A^{m \times n}$, which resembles the familiar statement for $m = n = 1$.

The content of this article is based on the bachelor dissertation [25] of the second-named author. The first-named author gratefully acknowledges partial support by the Austrian Science Fund (FWF) under project number P33878.

D. Dürrschnabel and D. López Rodríguez (Eds.): ICFCA 2023, LNAI 13934, pp. 109–126, 2023.
https://doi.org/10.1007/978-3-031-35949-1_8

Condition (†) defines a binary relation $\perp$ on the set $\mathcal{O}_A = \bigcup_{n \in \mathbb{N}_+} A^{A^n}$ of all finitary operations on A and as such an (infinite) context $\mathbb{K}_c = (\mathcal{O}_A, \mathcal{O}_A, \perp)$ that we refer to as the *context of commutation*. Since $\perp$ is a symmetric relation, both derivation operators on $\mathbb{K}_c$ are identical: for any $F \subseteq \mathcal{O}_A$ they compute $F^* := \{g \in \mathcal{O}_A \mid \forall f \in F \colon f \perp g\}$, the *centraliser* of F. Therefore, the systems of intents and extents of $\mathbb{K}_c$ each coincide with the set $\mathcal{C}_A = \mathfrak{I}(\mathbb{K}_c) = \mathfrak{E}(\mathbb{K}_c)$ of all so-called *centraliser clones* on A, i.e., with all those $F \subseteq \mathcal{O}_A$ satisfying $F^{**} = F$ (such sets of operations are also called *bicentrally closed*).

Commuting functions have found applications in aggregation theory in relation with the task of merging (aggregating) matrix data into a single value, cf. [2, 4,9,12,13] for more information. In particular, sets F of aggregation functions such that $F \subseteq F^*$, called *bisymmetric* in aggregation theory, have been subject to investigation, see [9,10]; in algebra such sets are simply called *self-commuting* and the algebras they induce are said to be *entropic*, generalising the classic notion of *mediality* for groupoids, quasigroups and semigroups, cf., e.g., [15,23].

For finite sets A, the lattice $\mathcal{C}_A$ of all centralisers is finite [8, Corollary 4], but even for small sets A its structure is quite complicated and still not very well understood. Thus, the unary parts $M := F^{*(1)} = \{s \in A^A \mid \forall f \in F \colon f \perp s\}$ of centralisers have received special attention in recent work, see, e.g. [2,3,6,11,16–21]. They are the intents of the (still infinite) subcontext[1] $\mathbb{K}_1 = (\mathcal{O}_A, A^A, \perp)$ of $\mathbb{K}_c$, and since they always form transformation monoids, they have been termed *centralising monoids*. Clearly, the representation of a centralising monoid M in the form $M = F^{*(1)}$ is not unique, and any set $F \subseteq \mathcal{O}_A$ satisfying this condition is called a *witness* of M. It is, however, immediate that the corresponding extent $M^* = F^{*(1)*} \supseteq F$ is always a maximal witness of M under set inclusion.

For $f \colon A^n \to A$ and $s \colon A \to A$, condition (†) specialises to the requirement

$$f(s(x_1), \dots, g(x_n)) = s(f(x_1, \dots, x_n)) \tag{$\ddagger$}$$

for all $x_1, \dots, x_n \in A$, which precisely expresses that s is an endomorphism of the algebra $\langle A; f \rangle$. Therefore, for $F \subseteq \mathcal{O}_A$ the functions $s \in F^{*(1)}$ are exactly the endomorphisms of the algebra $\langle A; F \rangle$, i.e., $F^{*(1)} = \mathrm{End}\langle A; F \rangle$. In other words, this means that for fixed A the (dually ordered) intent lattice of $\mathbb{K}_1$ is precisely the lattice of all possible endomorphism monoids of any possible algebraic structure on A, ordered by set inclusion. By inspecting the lattice, one may locate endomorphism monoids that are so-called 'cores', which is relevant for determining the complexity of solving systems of equations over algebras.

This article is concerned with the task of computationally enumerating all centralising monoids on a three-element set A, of giving witnesses for them in an automated fashion and of counting them. Rephrased in terms of algebraic structures, this means determining the elements and cardinality of the lattice of

---

[1] For brevity we shall forego restricting the general commutation relation by intersection with $\mathcal{O}_A \cap A^A$ even though this would be required to be meticulously correct.

endomorphism monoids on a three-element set, and, for each monoid, presenting an algebra such that the monoid is its endomorphism monoid.

For $|A| \leq 2$, *every* transformation monoid appears as the endomorphism monoid of some algebra. Machida and Rosenberg in [16] initiated a systematic study of centralising monoids (at the time called endoprimal monoids) on finite sets with a description of three co-atoms in the lattice of centralising monoids on $\{0, 1, 2\}$. This investigation was continued with the witness lemma from [17], giving a simple to check characterisation of when a set $F \subseteq \mathcal{O}_A$ is a witness of a centralising monoid M on some set A. With the help of the witness lemma, Machida and Rosenberg determined (by hand) all 51 centralising monoids on $A = \{0, 1, 2\}$ witnessed by unary functions, that is, the set of intents of the sub-context $(A^A, A^A, \perp)$ of $\mathbb{K}_1$. Moreover, Theorem 2.5 from [17] states the fundamental fact that every centralising monoid on a finite set A can be described by a finite witness, and a more refined version of the proof of [17, Theorem 2.5] will also be central for our investigations. In [18] it was observed that each maximal centralising monoid, i.e., a co-atom in the lattice under inclusion, can always be witnessed by a single minimal function appearing in the classification of minimal clones. Exploiting this connection for $|A| = 3$, precisely 10 maximal centralising monoids and corresponding singleton witnesses were determined in [18]. Extending these arguments, in [19] a strategy to determine all centralising monoids on $\{0, 1, 2\}$ was presented, and it was reported that their total number amounts to 192, splitting into 48 conjugacy types [19, Table 3]. Unfortunately, due to page restrictions, [19] does not contain complete details on how the monoids were obtained, nor a full list of all the monoids or any of their witnesses. The idea behind the proof is to consider all possible submonoids of each of the ten maximal centralising monoids and to sort out those that fail to be centralising. Failures are demonstrated by showing that a submonoid M is not closed under the intent-closure of $\mathbb{K}_1$, i.e., that $M^{**(1)} \supsetneq M$; the remaining ones are verified to be centralising by constructing witnesses, either directly, or by intersecting other already known centralising monoids. These steps are also outlined in the extended version [21] of [19], where tables with all 192 centralising monoids on $\{0, 1, 2\}$ can be found[2]. Both papers contain a few exemplary applications of the method sketched above, however, many details of how the 192 monoids on $\{0, 1, 2\}$ were computed and why they form a complete list are not given, neither in [19], nor in [21]. In particular, a complete step by step proof, as well as a list of 192 witnesses that would at least allow for straightforward checking of the given 192 monoids is still lacking. We note that, in theory, one could use the corresponding extent M^* as a maximal witness for each listed monoid M, but as M^* is infinite, it does not lend itself well to an automated check.

We are therefore interested in a computer-supported, mostly automated verification of Rosenberg and Machida's classification of all centralising monoids on $A = \{0, 1, 2\}$. For any finite set A, this is, in principle, a finite task, since there can be at most as many intents of $\mathbb{K}_1$ as there are subsets of A^A, but this naive upper bound $2^{|A|^{|A|}}$ already for $|A| = 3$ gives $2^{27} = 134\,217\,728$, which is every-

---

[2] Based on Sect. 4, our computations [5] will show that these are actually correct.

thing but small.[3] However, as the number of intents (and thus concepts) of $\mathbb{K}_1$ is finite, only a finite (in view of the number 192 from [21] possibly very small) portion of the infinite set $\mathcal{O}_A$ of objects of this context will actually be needed to discern all concepts. Still, a priori, it is not altogether clear where to make the cut, and finding a suitable division line will be a key challenge that we solve in this paper. With a little analysis (cf. [22, Theorem 3.1]) one can see—and we will also observe this as a consequence of our Theorem 14—that functions of arity $|A|$ suffice as objects, that is, the finite subcontext $\mathbb{K}_{|A|,1} = \left(A^{A^{|A|}}, A^A, \perp\right)$ has the same set of intents as $\mathbb{K}_1$, i.e., it describes all centralising monoids on A. However, for $|A| = 3$, this still gives 3^{27} ternary operations as objects, which is even more unwieldy than 2^{27} and certainly too large in order to enumerate all concepts of $\mathbb{K}_{3,1}$ with basic formal concept analysis tools. In fact, already storage of $\mathbb{K}_{3,1}$ would become problematic, even if each cross in the table only took one bit of memory.

We will call a set $W \subseteq \mathcal{O}_A$ *witness-complete* if $\mathfrak{I}(\mathbb{K}_1) = \mathfrak{I}\left(W, A^A, \perp\right)$, that is, if every centralising monoid M on A possesses a witness $F \subseteq W$. Clearly, every superset of a witness-complete set $W \subseteq \mathcal{O}_A$ remains witness-complete. In the terminology of formal concept analysis, witness-complete sets W correspond to specific *dense subcontexts* of $\mathbb{K}_1$ where the attribute set is not restricted. By results of [22], for finite A the object set $W_A = A^{A^{|A|}}$ is always finite and witness-complete. However, the previous discussion makes clear that already for $|A| = 3$ we need to be more precise and determine a much smaller witness-complete set in order to be able to enumerate the intents of $\mathbb{K}_1$ in a reasonable amount of time. This is what we will achieve by a careful analysis of the commutation relation of $\mathbb{K}_1$ in Sect. 3. We then describe how to implement our approach in practice, in order to obtain an object-clarified subcontext $\mathbb{K}'$ of $(W_A, A^A, \perp)$, where W_A is still witness-complete and, for the case $|A| = 3$, contains only 183 functions as objects. Moreover, for the carrier set $\{0, 1, 2\}$, we will be able to present a row-reduced subcontext $\mathbb{K}''$ of $\mathbb{K}'$ with only 51 objects, from which one obtains the standard context of $\mathbb{K}_1$ by only deleting two redundant attributes. By computing the concept lattice of $\mathbb{K}'$ (or $\mathbb{K}''$), we will be able to verify the total number 192 of centralising monoids on $\{0, 1, 2\}$ given by Machida and Rosenberg in [19] and to check that there are 48 conjugacy types of them. Moreover, the concepts will allow us to present the list of monoids including their witnesses, which turns at least any future verification of the figure 192 as a lower bound into an immediate task. To infer the completeness of this list of 192 monoids, one still has to trust the correctness of Machida and Rosenberg's or our computations. In order to facilitate that, we shall deposit the code we used and supplementary data in [5].

---

[3] Since our purpose is to verify the results of [19], we are not going to rely on the bound 192 for the number of concepts.

2   Preliminaries

We first introduce some notation regarding operations that has not already been explained in the introduction. Furthermore, we recollect some known facts that will be necessary to study the context $\mathbb{K}_1$.

We write $\mathbb{N} = \{0, 1, 2, \ldots\}$ for the set of natural numbers and abbreviate $\mathbb{N}_+ := \mathbb{N} \setminus \{0\}$. We denote the set of functions from B to A as A^B. If $f \colon C \to B$ and $g \colon B \to A$, we understand their composition from right to left, that is, $g \circ f \colon C \to A$ maps any $x \in C$ to $g(f(x)) \in A$. For $n \in \mathbb{N}$, an n-ary operation on A is just a map $f \colon A^n \to A$; the parameter n is referred to as the arity of f. For a unary constant operation with value $a \in A$ we use the notation $c_a \colon A \to A$. We collect all finitary (non-nullary) operations on A in the set $\mathcal{O}_A = \bigcup_{n \in \mathbb{N}_+} A^{A^n}$. For a set $F \subseteq \mathcal{O}_A$ and $n \in \mathbb{N}$, we denote by $F^{(n)} := F \cap A^{A^n}$ its n-ary part, in particular we have $\mathcal{O}_A^{(n)} = A^{A^n}$. Moreover, we set $F^{(\leq n)} := \bigcup_{j=0}^n F^{(j)}$. With this notation, the context $\mathbb{K}_1$ becomes $(\mathcal{O}_A, \mathcal{O}_A^{(1)}, \bot)$, and for $|A| = k \in \mathbb{N}$ we have $\mathbb{K}_{k,1} = (\mathcal{O}_A^{(k)}, \mathcal{O}_A^{(1)}, \bot)$.

We also understand tuples $\boldsymbol{x} \in A^n$ as maps $\boldsymbol{x} = (\boldsymbol{x}(0), \ldots, \boldsymbol{x}(n-1))$ from $n = \{0, \ldots, n-1\}$ into A; given $0 \leq i < n$, we often write $\boldsymbol{x}_i$ for $\boldsymbol{x}(i)$. We are thus able to compose tuples with maps $h \colon A \to B$ from the outside, giving $h \circ \boldsymbol{x} = (h(\boldsymbol{x}_0), \ldots, h(\boldsymbol{x}_{n-1})) \in B^n$, or with maps $\alpha \colon I \to n$ from the inside, resulting in $\boldsymbol{x} \circ \alpha = (\boldsymbol{x}_{\alpha(i)})_{i \in I} \in A^I$. In this way, if $n, m \in \mathbb{N}$, $f \colon A^n \to A$ and $\alpha \colon n \to m$, we can define the minor of f induced by α as the operation $\delta_\alpha f \colon A^m \to A$ given by $\delta_\alpha f(\boldsymbol{x}) := f(\boldsymbol{x} \circ \alpha)$ for all $\boldsymbol{x} \in A^m$. Computing minors is functorial, that is, if we take another index map $\beta \colon m \to \ell$ with $\ell \in \mathbb{N}$, then we have $\delta_\beta(\delta_\alpha f)(\boldsymbol{x}) = \delta_\alpha f(\boldsymbol{x} \circ \beta) = f(\boldsymbol{x} \circ \beta \circ \alpha) = \delta_{\beta \circ \alpha} f(\boldsymbol{x})$ for all $\boldsymbol{x} \in A^\ell$, implying $\delta_\beta(\delta_\alpha f) = \delta_{\beta \circ \alpha} f$, i.e., $\delta_\beta \circ \delta_\alpha = \delta_{\beta \circ \alpha}$. Moreover, $\delta_{\mathrm{id}_n} f = f$, i.e., $\delta_{\mathrm{id}_n} = \mathrm{id}_{\mathcal{O}_A^{(n)}}$. If α is non-injective, arguments of f get identified by $\delta_\alpha f$; if α is non-surjective, $\delta_\alpha f$ will incur fictitious arguments. We will call those minors of f where α is surjective identification minors (or just variable identifications) of f, and we will speak of proper variable identifications if α is surjective, but not injective. The former notion necessitates that $n \geq m$ for $\alpha \colon n \to m$, and the latter that $m < n$. A particulary relevant example of a proper variable identification is given for $0 \leq i < j < n$ by the map $\alpha_{ij} \colon n \to n-1$, defined as $\alpha_{ij}(\ell) := \ell$ if $0 \leq \ell \leq j-1$, $\alpha_{ij}(j) := i$ and $\alpha_{ij}(\ell) := \ell-1$ if $j+1 \leq \ell < n$. We will call $\Delta_{ij} f := \delta_{\alpha_{ij}} f$ the (i,j)-variable identification (minor) of f; let us note that $\Delta_{ij} f(x_0, \ldots, x_{n-2}) = f(x_0, \ldots, x_{j-1}, x_i, x_j, \ldots, x_{n-2})$ for all $x_0, \ldots, x_{n-2} \in A$. We will say that a set $F \subseteq \mathcal{O}_A$ is closed under (proper) variable identifications if it contains with each member $f \in F$ also any (proper) identification minor of f. We will use an analogous definition for $F \subseteq \mathcal{O}_A$ being closed under taking general minors.

For $n \in \mathbb{N}_+$ and $0 \le i < n$, we define the i-th n-ary *projection* to be the operation $e_i^{(n)} \in \mathcal{O}_A^{(n)}$ given by $e_i^{(n)}(x_0, \ldots, x_{n-1}) := x_i$ for every $(x_0, \ldots, x_{n-1}) \in A^n$. The operation $\mathrm{id}_A = e_0^{(1)}$ is the identity operation on A. For $n, m \in \mathbb{N}_+$ and $f \in \mathcal{O}_A^{(n)}$, $g_0, \ldots, g_{n-1} \in \mathcal{O}_A^{(m)}$, we define their *composition* $h := f \circ (g_0, \ldots, g_{n-1})$ as the operation $h \in \mathcal{O}_A^{(m)}$ given for $\boldsymbol{x} \in A^m$ by $h(\boldsymbol{x}) := f(g_0(\boldsymbol{x}), \ldots, g_{n-1}(\boldsymbol{x}))$. A *clone* on A is any set $F \subseteq \mathcal{O}_A$ such that $e_i^{(n)} \in F$ for every $n \in \mathbb{N}_+$ and $0 \le i < n$ and F is closed under composition, that is, we require $f \circ (g_0, \ldots, g_{n-1}) \in F$ whenever $f, g_0, \ldots, g_{n-1} \in F$ are of matching arities. The following basic lemma about clones is evident.

Lemma 1. *If $F \subseteq \mathcal{O}_A$ is a clone, then it is closed under taking minors.*

Proof. For $n, m \in \mathbb{N}_+$, $f \in F^{(n)}$, $\alpha: n \to m$, the minor $\delta_\alpha f$ can be expressed as a composition of f and projections: $\delta_\alpha f = f \circ \left(e_{\alpha(0)}^{(m)}, \ldots, e_{\alpha(n-1)}^{(m)}\right) \in F$. $\square$

We shall need the fact that all centralisers are clones. For this we will introduce the notion of polymorphism of a set of relations. For $m \in \mathbb{N}$, an m-ary *relation* on A is just any subset $\varrho \subseteq A^m$. The set of all relations on A of any possible arity $m \in \mathbb{N}_+$ is denoted as $\mathcal{R}_A$. We say that an operation $f \in \mathcal{O}_A^{(n)}$ *preserves* ϱ if for all tuples $\boldsymbol{r}_1, \ldots, \boldsymbol{r}_n \in \varrho$ it is the case that the m-tuple $f \circ (\boldsymbol{r}_1, \ldots, \boldsymbol{r}_n)$ obtained by component-wise application of f also belongs to ϱ. We symbolise the truth of this fact by $f \rhd \varrho$. The set of *polymorphisms* of a set $Q \subseteq \mathcal{R}_A$ of relations is then defined as $\mathrm{Pol}\, Q := \{ f \in \mathcal{O}_A \mid \forall \varrho \in Q \colon f \rhd \varrho \}$, i.e., it is one derivation operator of the context $(\mathcal{O}_A, \mathcal{R}_A, \rhd)$. It is well known that the extents, that is, all polymorphism sets, are clones.

Lemma 2 ([24, 1.1.15 Satz]). *For any $Q \subseteq \mathcal{R}_A$ the set $\mathrm{Pol}\, Q$ is a clone on A.*

We can represent centralisers as polymorphism clones by understanding the functions that are centralised as relations via their graphs. That is, for $n \in \mathbb{N}_+$ and $f \in \mathcal{O}_A^{(n)}$ we define $f^\bullet := \{ (x_0, \ldots, x_n) \in A^{n+1} \mid f(x_0, \ldots, x_{n-1}) = x_n \}$ to be the graph of f, and we set $F^\bullet := \{ f^\bullet \mid f \in F \}$ for any $F \subseteq \mathcal{O}_A$. Then we have the following (also well-known) connection to the centraliser of F.

Lemma 3 ([24, Üb. 2.10, p. 75]). *For every $F \subseteq \mathcal{O}_A$ we have $F^* = \mathrm{Pol}\, F^\bullet$.*

As centralisers are the central topic of this article, we supply a short proof.

Proof Let $m, n \in \mathbb{N}$, $g \in \mathcal{O}_A^{(n)}$ and $f \in F^{(m)}$. If $g \in \operatorname{Pol} F^\bullet$, then $g \rhd f^\bullet$ and thus for any matrix $(x_{ij})_{1 \leq i \leq m, 1 \leq j \leq n} \in A^{m \times n}$ we have

$$\begin{pmatrix} g(x_{11}, \ldots, x_{1n}) \\ \vdots \\ g(x_{m1}, \ldots, x_{mn}) \\ g(f(x_{11}, \ldots, x_{m1}), \ldots, f(x_{1n}, \ldots, x_{mn})) \end{pmatrix} \in f^\bullet,$$

since for each $1 \leq j \leq n$ the column $(x_{1j}, \ldots, x_{mj}, f(x_{1j}, \ldots, x_{mj}))$ belongs to $f^\bullet$. The latter is equivalent to $g(f(x_{11}, \ldots, x_{m1}), \ldots, f(x_{1n}, \ldots, x_{mn}))$ being equal to $f(g(x_{11}, \ldots, x_{1n}), \ldots, g(x_{m1}, \ldots, x_{mn}))$, whence $g \perp f$. For the converse implication, we start from $g \in F^*$, implying $g \perp f$ and thus entailing the truth of the previous equality. Therefore, g indeed maps any n tuples in $f^\bullet$, which are given as the columns of an $(m \times n)$-matrix extended by an additional row of column values, into $f^\bullet$. Hence g preserves $f^\bullet$.   □

Corollary 4. *For any $F \subseteq \mathcal{O}_A$, the centraliser F^* is a clone, and is hence closed under taking minors, in particular, it is closed under variable identifications.*

Proof. By Lemma 3, $F^* = \operatorname{Pol} F^\bullet$ is a polymorphism set, which by Lemma 2 is a clone and therefore closed under taking minors, see Lemma 1.   □

Another direct consequence of Corollary 4 is that each centralising monoid $F^{*(1)}$ is a transformation monoid, for it is the unary part of the clone F^* and thus contains id_A and is closed under composition. With respect to centralising monoids we also cite the following result from [22], which we will improve in Theorem 14.

Lemma 5 ([22, **Theorem 3.1**]). *Any centralising monoid M on a finite set A has a finite witness $F \subseteq \mathcal{O}_A^{(\leq k)}$ consisting of functions of arity at most $k = |A|$.*

We also need a description of centralisers of constants, which is easy to verify.

Lemma 6 ([18, **Lemma 5.2**], [6, **Lemma 5**]). *For any set A and every $a \in A$ we have $\{c_a\}^* = \operatorname{Pol} \{\{a\}\}$, i.e., for $f \in \mathcal{O}_A$ we have $f \perp c_a$ iff $f(a, \ldots, a) = a$.*

3   Finding Witness-Complete Sets

The following simple observation lies at the basis of our construction of witness-complete sets.

Lemma 7. *Let $n \in \mathbb{N}$, $0 \leq i < j < n$, $f \in \mathcal{O}_A^{(n)}$ and $s \in \mathcal{O}_A^{(1)}$. Moreover, let $a = (a_0, \ldots, a_{n-1}) \in A^n$ be such that $a_i = a_j$ and $s(f(a)) \neq f(s \circ a)$. Then $\Delta_{ij} f$ has arity $n - 1$ and does not belong to $\{s\}^*$.*

Proof. Since $a_j = a_i$ and thus also $s(a_j) = s(a_i)$, we have

$$
\begin{aligned}
s(\Delta_{ij} f(a_0, \ldots, a_{j-1}, a_{j+1}, \ldots, a_n)) &= s(f(a_0, \ldots, a_{j-1}, a_i, a_{j+1}, \ldots, a_n)) \\
&= s(f(a_0, \ldots, a_{j-1}, a_j, a_{j+1}, \ldots, a_n)) \\
&\neq f(s(a_0), \ldots, s(a_{j-1}), s(a_j), s(a_{j+1}), \ldots, s(a_n)) \\
&= f(s(a_0), \ldots, s(a_{j-1}), s(a_i), s(a_{j+1}), \ldots, s(a_n)) \\
&= \Delta_{ij} f(s(a_0), \ldots, s(a_{j-1}), s(a_{j+1}), \ldots, s(a_n)),
\end{aligned}
$$

whence $\Delta_{ij} f$ fails to commute with s on the tuple $(a_0, \ldots, a_{j-1}, a_{j+1}, \ldots, a_{n-1})$ in A^{n-1}. □

The following result collects basic properties of non-commuting operations of least possible arity.

Proposition 8. *Let A be a finite set, $s \in \mathcal{O}_A^{(1)}$, $F \subseteq \mathcal{O}_A$ be closed under proper variable identifications and $f \in F \setminus \{s\}^*$ be of minimal arity $n \in \mathbb{N}_+$. Then the following facts hold:*

(i) *Every proper variable identification of f belongs to $\{s\}^*$.*
(ii) *The arity n is bounded above by $|A|$, that is, $n \leq |A|$.*
(iii) *For every non-injective tuple $\boldsymbol{a} \in A^n$ we have $s(f(\boldsymbol{a})) = f(s \circ \boldsymbol{a})$.*
(iv) *There is some injective tuple $\boldsymbol{b} \in A^n$ for which $s(f(\boldsymbol{b})) \neq f(s \circ \boldsymbol{b})$.*

Proof. Let $m < n$ and $\alpha \colon n \to m$ be any surjective (and necessarily non-injective) map. Since F is closed under proper variable identifications and $f \in F$, we have $\delta_\alpha f \in F^{(m)}$. As f has least possible arity in $F \setminus \{s\}^*$ and $m < n$, the condition $\delta_\alpha f \notin \{s\}^*$ must fail, for otherwise $\delta_\alpha f$ would belong to $F \setminus \{s\}^*$. Hence, $\delta_\alpha f \in \{s\}^*$, proving (i).

If $\boldsymbol{a} = (a_0, \ldots, a_{n-1})$ is any non-injective tuple in A^n, then there are indices $0 \leq i < j < n$ such that $a_i = a_j$. As F is closed under proper variable identifications, $\Delta_{ij} f \in F^{(n-1)}$. If $s(f(\boldsymbol{a})) \neq f(s \circ \boldsymbol{a})$, then, by Lemma 7, we have $\Delta_{ij} f \in F^{(n-1)} \setminus \{s\}^*$, in contradiction to the minimality of n. Hence, we have (iii).

Since $f \notin \{s\}^*$ there must be some tuple $\boldsymbol{b} \in A^n$ for which $s(f(\boldsymbol{b})) \neq f(s \circ \boldsymbol{b})$. By (iii), this tuple cannot be non-injective, thus $\boldsymbol{b}$ is injective and (iv) holds.

If n were larger than $|A|$, then every n-tuple would be non-injective, hence a tuple $\boldsymbol{b}$ as promised in (iv) could not exist. Thus, we have $n \leq |A|$, i.e., (ii). □

In order to sharpen Lemma 5, we shall rely on the following result.

Lemma 9. *For any finite set A and each $s \in \mathcal{O}_A^{(1)}$, there exists $n_s \in \mathbb{N}$, subject to $1 \leq n_s \leq |A| := k$, such that for every $F \subseteq \mathcal{O}_A$ that is closed under proper variable identifications, the implication $F \not\subseteq \{s\}^* \implies F^{(\leq n_s)} \not\subseteq \{s\}^*$ holds. If, moreover, F is closed under taking minors, then $F \not\subseteq \{s\}^*$ even implies $F^{(n_s)} \not\subseteq \{s\}^*$.*

Proof. Let $s \in \mathcal{O}_A^{(1)}$ and an identification closed set $F \subseteq \mathcal{O}_A$ be given, and assume that $F \not\subseteq \{s\}^*$. This means there exists some operation $f \in F \setminus \{s\}^*$; let us choose such a function f of minimum possible arity $n_{s,F} \in \mathbb{N}_+$. From Proposition 8(ii) we have $n_{s,F} \leq k$, wherefore $I_s := \{n_{s,G} \mid G \not\subseteq \{s\}^*$ closed w.r.t. proper variable identifications$\}$ is a nonempty subset of $\{1, \ldots, k\}$, and hence $n_s := \max I_s$ satisfies $1 \leq n_s \leq k$. As $n_{s,F} \in I_s$, we infer $n_{s,F} \leq n_s \leq k$. Thus, $f \in F^{(n_{s,F})} \setminus \{s\}^* \subseteq F^{(\leq n_s)} \setminus \{s\}^*$, and therefore $F^{(\leq n_s)} \not\subseteq \{s\}^*$. If we assume F to be closed under minors in general, then we can use the fictitious minor $g := \delta_\alpha f \in F$, where $\alpha \colon n_{s,F} \hookrightarrow n_s$ is the identical embedding, to prove that $g \in F^{(n_s)} \setminus \{s\}^*$, i.e., $F^{(n_s)} \not\subseteq \{s\}^*$. Namely, by using the identification map $\beta \colon n_s \to n_{s,F}$ given as $\beta(\ell) := \ell$ for indices $0 \leq \ell < n_{s,F}$ and $\beta(\ell) := n_{s,F} - 1$ else, we have $\beta(\alpha(\ell)) = \beta(\ell) = \ell$ for $0 \leq \ell < n_{s,F}$, and thus $\beta \circ \alpha = \mathrm{id}_{n_{s,F}}$. By functoriality of taking minors, we conclude $\delta_\beta g = \delta_\beta(\delta_\alpha f) = \delta_{\beta \circ \alpha} f = \delta_{\mathrm{id}_{n_{s,F}}} f = f \notin \{s\}^*$, which by Corollary 4 implies that indeed $g \notin \{s\}^*$, for otherwise the centraliser $\{s\}^*$ would have to contain the minor $\delta_\beta g = f$. $\qquad\square$

As an immediate consequence we may uniformly take $n_s = |A|$ for *every* $s \in \mathcal{O}_A^{(1)}$ to satisfy the implication of Lemma 9 on a finite set A.

Corollary 10. *For any finite set A of size $|A| = k$, each $s \in \mathcal{O}_A^{(1)}$ and every $F \subseteq \mathcal{O}_A$ satisfying $F \not\subseteq \{s\}^*$ and being closed under proper variable identifications, we have $F^{(\leq k)} \not\subseteq \{s\}^*$.*

Proof. Given $s \in \mathcal{O}_A^{(1)}$, take $n_s \leq k$ constructed in Lemma 9. Then any function set $F \not\subseteq \{s\}^*$ that is closed under proper variable identifications fulfils, again by Lemma 9, $\emptyset \neq F^{(\leq n_s)} \setminus \{s\}^* \subseteq F^{(\leq k)} \setminus \{s\}^*$, as required. $\qquad\square$

For any finite set A of size $|A| = k$ and every $s \in \mathcal{O}_A^{(1)}$, Lemma 9 guarantees the existence of a smallest arity $n_s \in \{1, \ldots, k\}$ such that every $F \not\subseteq \{s\}^*$ that is closed under variable identifications satisfies $F^{(\leq n_s)} \not\subseteq \{s\}^*$. We shall mainly exploit this implication for very specific types of clones, in fact, specific centraliser clones, and it may well be that for some $s \in \mathcal{O}_A^{(1)}$ there exists a smaller arity n_s which satisfies the implications claimed by Lemma 9 under this restriction. We shall not explore finding the smallest possible value of n_s in detail, except for simple but important cases.

Lemma 11. *If $s = \mathrm{id}_A$ or $s = c_a$ for some $a \in A$, then $F \not\subseteq \{s\}^*$ implies $F^{(1)} \not\subseteq \{s\}^*$ for all $F \subseteq \mathcal{O}_A$ that are closed under proper variable identifications.*

Proof. If $s = \mathrm{id}_A$ the implication holds since the assumption $F \not\subseteq \{s\}^* = \mathcal{O}_A$ is never satisfied. Let us hence consider $s = c_a$ with $a \in A$. By Lemma 6 we know $\{c_a\}^* = \mathrm{Pol}\{\{a\}\}$, thus for any $F \subseteq \mathcal{O}_A$ we have $F \setminus \{c_a\}^* \neq \emptyset$ iff there is $f \in F$ such that $f(a, \ldots, a) \neq a$, i.e., iff there is $f \in F$ for which $\delta_\alpha f(a) \neq a$ holds with α being the unique map into $1 = \{0\}$. Assuming that F is closed under proper variable identifications, we have $\delta_\alpha f \in F^{(1)}$, hence the previous condition implies that $F^{(1)} \setminus \{c_a\}^* \neq \emptyset$, or equivalently $F^{(1)} \not\subseteq \{c_a\}^*$. $\qquad\square$

In our construction of witnesses for finite A we shall follow the proof outlined in [17, Theorem 2.5], but we treat each $s \in \mathcal{O}_A^{(1)}$ individually and exploit minimal arities of counterexamples to commutation, thus being able to obtain a smaller witness-complete set than $\mathcal{O}_A^{(|A|)}$. Keeping Proposition 8(i) in mind, we first define the following set $\Phi_{s,n}$ for each $s \in \mathcal{O}_A^{(1)}$ and $n \in \mathbb{N}_+$.

Definition 12. *For $s \in \mathcal{O}_A^{(1)}$ and each $n \in \mathbb{N}_+$, we set*

$$\Phi_{s,n} := \left\{ f \in \mathcal{O}_A^{(n)} \mid \forall 0 \le m < n \, \forall surjective \; \alpha \colon n \twoheadrightarrow m \colon \quad \delta_\alpha f \in \{s\}^* \right\}.$$

For computational purposes, the following characterisation is useful.

Lemma 13. *Given any $s \in \mathcal{O}_A^{(1)}$ and $n \in \mathbb{N}_+$ we have the equivalent description*

$$\Phi_{s,n} = \left\{ f \in \mathcal{O}_A^{(n)} \mid \forall 0 \le i < j < n \colon \Delta_{ij} f \in \{s\}^* \right\}$$
$$= \left\{ f \in \mathcal{O}_A^{(n)} \mid \forall 0 \le m < n \, \forall \alpha \colon n \to m \colon \delta_\alpha f \in \{s\}^* \right\}.$$

Proof. If $f \in \Phi_{s,n}$ and $0 \le i < j < n$, then clearly $\Delta_{ij} f = \delta_{\alpha_{ij}} \in \{s\}^*$ because $\alpha_{ij} \colon n \twoheadrightarrow n-1$ is surjective and non-injective due to $i < j$. For the second inclusion assume that $f \in \mathcal{O}_A^{(n)}$ satisfies $\Delta_{ij} f \in \{s\}^*$ for all $0 \le i < j < n$ and consider any $m < n$ and any $\alpha \colon n \to m$. As $m < n$, the map α cannot be injective, hence there are $0 \le i < j < n$ for which $\alpha(i) = \alpha(j)$. By our assumption we have $\Delta_{ij} f \in \{s\}^*$. We define $\beta \colon n-1 \to m$ by $\beta(\ell) := \alpha(\ell)$ if $0 \le \ell < j$ and $\beta(\ell) := \alpha(\ell+1)$ if $j \le \ell \le n-2$. Hence, $\beta(\alpha_{ij}(\ell)) = \beta(\ell) = \alpha(\ell)$ if $0 \le \ell < j$, $\beta(\alpha_{ij}(j)) = \beta(i) = \alpha(i) = \alpha(j)$, and $\beta(\alpha_{ij}(\ell)) = \beta(\ell-1) = \alpha(\ell-1+1) = \alpha(\ell)$ for all $j < \ell < n$. Thus, we have $\beta \circ \alpha_{ij} = \alpha$, and, by functoriality of taking minors, we infer $\delta_\beta \circ \delta_{\alpha_{ij}} = \delta_{\beta \circ \alpha_{ij}} = \delta_\alpha$, that is, $\delta_\alpha f = \delta_\beta(\Delta_{ij} f) \in \{s\}^*$ because $\Delta_{ij} f \in \{s\}^*$ and the centraliser $\{s\}^*$ is closed under minors by Corollary 4. The remaining containment relation in $\Phi_{s,n}$ obviously follows by specialisation. $\square$

We now provide the theoretical backbone for constructing smaller witness-complete sets. The result is obtained by refining the proof techniques used in [17, Theorem 2.5] and [22, Theorem 3.1].

Theorem 14. *Let A be a finite set of size $|A| = k$, and let $(n_s)_{s \in \mathcal{O}_A^{(1)} \setminus \{\mathrm{id}_A\}}$ be any tuple of integers $1 \le n_s \le k$ such that for each $s \in \mathcal{O}_A^{(1)} \setminus \{\mathrm{id}_A\}$ the implication claimed by Lemma 9 is satisfied. Setting $\Phi_s := \bigcup_{1 \le \ell \le n_s} \Phi_{s,\ell}$ for each $s \ne \mathrm{id}_A$, the set $\Phi := \bigcup_{s \in \mathcal{O}_A^{(1)} \setminus \{\mathrm{id}_A\}} \Phi_s$ is witness-complete w.r.t. $\mathbb{K}_1$.*

Proof. Let $M \subseteq \mathcal{O}_A^{(1)}$ be a centralising monoid, i.e., an intent of $\mathbb{K}_1$. Thus, $M = M^{**(1)}$ is intent-closed, hence $M = F^{*(1)}$ for the witness $F = M^*$. We observe from Corollary 4 that $F = M^*$ is closed under arbitrary minors, in particular under (proper) variable identifications. We want to reduce F to a subset $G \subseteq F \cap \Phi$ that still witnesses M. Clearly, we have $G^{*(1)} \supseteq F^{*(1)} = M$ for any $G \subseteq F$; the crucial point is to keep enough functions of $F \cap \Phi$ in G

such that this inclusion does not become proper. In other words, we need to ensure that $G^{*(1)} \subseteq M$, or, equivalently, that the complement of M in $\mathcal{O}_A^{(1)}$ has empty intersection with G^*. For this reason, we consider an arbitrary function $s \in \mathcal{O}_A^{(1)} \setminus M$; since $\mathrm{id}_A \in M$, we have $s \neq \mathrm{id}_A$. As $s \in \mathcal{O}_A^{(1)}$ but $s \notin M = F^{*(1)}$, there is some $f_s \in F = M^*$ where $f_s \perp s$ fails, i.e., $f_s \in F \setminus \{s\}^*$. Let us, in fact, choose $f_s \in F \setminus \{s\}^*$ of least possible arity $n \in \mathbb{N}_+$. From Proposition 8(i) we know that every proper variable identification of f_s belongs to $\{s\}^*$, i.e., that $f_s \in \Phi_{s,n}$. Since $f_s \in F \setminus \{s\}^*$ we have $F \not\subseteq \{s\}^*$; moreover, F is closed under (proper) variable identifications. Therefore, we can invoke Lemma 9 to conclude that $F^{(\leq n_s)} \not\subseteq \{s\}^*$, which means that there is some $g_s \in F \setminus \{s\}^*$ of arity $m \leq n_s$. Since by our choice f_s had minimum possible arity in $F \setminus \{s\}^*$, it follows that $n \leq m \leq n_s$. Therefore, we have $f_s \in \Phi_{s,n} \subseteq \Phi_s \subseteq \Phi$. Picking now one such f_s for each $s \neq \mathrm{id}_A$, we have found a subset $G = \left\{ f_s \mid s \in \mathcal{O}_A^{(1)} \setminus M \right\} \subseteq F \cap \Phi$, for which $\left(\mathcal{O}_A^{(1)} \setminus M \right) \cap G^* = \emptyset$ since every $s \in \mathcal{O}_A^{(1)} \setminus M$ does not commute with its chosen $f_s \in F \setminus \{s\}^*$. Thus $G^{*(1)} \subseteq M = F^{*(1)} \subseteq G^{*(1)}$ and $G \subseteq \Phi$ is a witness of M. $\qquad \square$

Corollary 10 allows us to instantiate all values n_s in Theorem 14 in the form $n_s = k = |A|$, wherefore $\Phi_s \subseteq \mathcal{O}_A^{(\leq k)}$ for each $s \neq \mathrm{id}_A$, and thus the theorem produces a witness-complete set $\Phi \subseteq \mathcal{O}_A^{(\leq k)}$. As every superset of a witness-complete set retains this property, we conclude that $\mathcal{O}_A^{(\leq k)}$ is witness-complete, i.e., Lemma 5 is a special case of Theorem 14.

For practical application we shall combine Theorem 14 with the idea of object-clarification. We say that objects $f, g \in \mathcal{O}_A$ of $\mathbb{K}_1$ are *equivalent in* $\mathbb{K}_1$, and denote this fact by $f \equiv g$, if they witness the same centralising monoid, that is, $\{f\}^{*(1)} = \{g\}^{*(1)}$. A transversal picks exactly one representative from each $\equiv$-class, and we will now proceed to show that any superset of a transversal of a witness-complete set is still suitable for our purpose.

Proposition 15. *Let A be a set, $W \subseteq \mathcal{O}_A$ be witness-complete w.r.t. $\mathbb{K}_1$ and let $T \subseteq \mathcal{O}_A$ be such that each $f \in W$ is equivalent to some $g \in T$; then T is also witness-complete.*

Proof. Let $M = F^{*(1)} \subseteq \mathcal{O}_A$ be a centralising monoid, witnessed by $F \subseteq W$. We define $G := \{g \in T \mid \exists f \in F \colon f \equiv g\} \subseteq T$, and claim that $G^{*(1)} = F^{*(1)} = M$. Since $F \subseteq W$, for each $f \in F$ there is some $g_f \in T$ such that $f \equiv g_f$, that is, $\{f\}^{*(1)} = \{g_f\}^{*(1)}$. By definition of G we have $g_f \in G$, wherefore we obtain $F^{*(1)} = \bigcap_{f \in F} \{f\}^{*(1)} = \bigcap_{f \in F} \{g_f\}^{*(1)} \subseteq \bigcap_{g \in G} \{g\}^{*(1)} = G^{*(1)}$. Conversely, each $g \in G$ is equivalent to some $f \in F$ by the definition of G. Hence, by a symmetric argument, we infer $G^{*(1)} \subseteq F^{*(1)}$, and thus $M = F^{*(1)} = G^{*(1)}$ with $G \subseteq T$. $\qquad \square$

4   Computing All Centralising Monoids on $\{0, 1, 2\}$

In principle, it is possible to apply the steps shown in this section to simplify $\mathbb{K}_1 = \left(\mathcal{O}_A, \mathcal{O}_A^{(1)}, \perp\right)$ for any finite set A. However, for $|A| \leq 2$ the problem is trivial, and in view of data from [1,6], it seems unlikely that the approach shown here will be feasible for $|A| \geq 4$ from a computational perspective. We shall therefore only discuss the case $A = \{0, 1, 2\} = 3$ in this section.

Let us define the integer sequence $(n_s)_{s \in \mathcal{O}_3^{(1)}}$ by $n_s := 1$ if $s = \mathrm{id}_3$ or $s = c_a$ with $0 \leq a < 3$, and $n_s := 3$ else. By Corollary 10 and Lemma 11, this sequence can be used as a parameter in Theorem 14 to produce an initial witness-complete set $\Phi = \bigcup_{s \in \mathcal{O}_3^{(1)} \setminus \{\mathrm{id}_3\}} \Phi_s$. Since $\Phi_{c_a} \subseteq \mathcal{O}_3^{(1)}$ and $\Phi_{s,n} \subseteq \mathcal{O}_3^{(\leq 2)}$ for $n \leq 2$ and any $s \in \mathcal{O}_3^{(1)}$, we have $\Phi \subseteq \mathcal{O}_3^{(\leq 2)} \cup \bigcup_{s \in \mathcal{U}_3} \Phi_{s,3}$ where $\mathcal{U}_3 := \mathcal{O}_3^{(1)} \setminus \{\mathrm{id}_3, c_0, c_1, c_2\}$. Let $\iota \colon 1 \hookrightarrow 2$ be the identical inclusion; we certainly have $u \equiv \delta_\iota u$ for any $u \in \mathcal{O}_3^{(1)}$, where $\delta_\iota u \in \mathcal{O}_3^{(2)}$ is u with a fictitious argument added. Thus, by Proposition 15, the set $T := \mathcal{O}_3^{(2)} \cup \bigcup_{s \in \mathcal{U}_3} \Phi_{s,3}$ is witness-complete. Our goal is to enumerate a superset of T, to clarify it and use it as objects of a subcontext of $\mathbb{K}_1$: For each $s \in \mathcal{U}_3$, we can easily enumerate all binary operations in $\{s\}^*$ by brute force, as $\left|\mathcal{O}_3^{(2)}\right| = 3^9 = 19\,683$. Then, for any $f \in \Phi_{s,3}$ where $s \in \mathcal{U}_3$, we know from Lemma 13 that $\Delta_{01} f, \Delta_{02} f, \Delta_{12} f \in \{s\}^*$. Moreover, for any $x \in A$ and $ij \in \{01, 02, 12\}$, we must have $f(x, x, x) = \Delta_{ij} f(x, x)$, that is, the functions $\Delta_{01} f, \Delta_{02} f, \Delta_{12} f$ must be 'compatible on the diagonal'. Thus we will be enumerating all f in a superset $\Psi_s \supseteq \Phi_{s,3}$ if we iterate over all possible combinations of $g_1, g_2, g_3 \in \{s\}^{*(2)}$ that are compatible on the diagonal, iterate over all functions $h \in A^S$ where $S = \left\{ (x, y, z) \in A^3 \mid x \neq y \neq z \neq x \right\}$ and define $f(x, y, z) := g_1(y, z)$ if $x = y$, $f(x, y, z) := g_2(x, y)$ if $x = z$, $f(x, y, z) := g_3(x, y)$ if $y = z$, and $f(x, y, z) := h(x, y, z)$ if $(x, y, z) \in S$. While enumerating the witness-complete superset $\mathcal{O}_3^{(2)} \cup \bigcup_{s \in \mathcal{U}_3} \Psi_s \supseteq T$, we can apply Proposition 15 on the fly by not storing a function f as an object a second time if its intent $\{f\}^{*(1)}$ has already been encountered before. This drastically reduces the amount of stored functions to a witness-complete subset $W \subseteq \mathcal{O}_3^{(2)} \cup \bigcup_{s \in \mathcal{U}_3} \Psi_s$ such that $|W| = 183$ and $\mathbb{K}' = \left(W, \mathcal{O}_3^{(1)}, \perp\right)$ is an object-clarified subcontext of $\mathbb{K}_1$ with identical set of intents. The context $\mathbb{K}'$ can be found in [25, Sections 7.2, 7.3].

We remark that the computation of $\mathbb{K}'$ for $|A| = 3$ is feasible for the following reasons. Brute-force enumeration of $\{s\}^{*(2)}$ for $s \in \mathcal{O}_3^{(1)}$ is not of any concern. The crucial point, however, lies in the ability to exploit Lemma 11 to reduce the union over all $s \in \mathcal{O}_3^{(1)} \setminus \{\mathrm{id}_3\}$ in Φ to the union over all $s \in \mathcal{U}_3$ in T. Namely, $\max_{s \in \mathcal{U}_3} \left|\{s\}^{*(2)}\right| = 162$, while a constant c_a commutes with all $3^8 = 6\,561$ binary operations b satisfying $b(a, a) = a$ (Lemma 6). Luckily, $|A^S| = 3^6 = 729$ is small. The core loop in the enumeration of W [5] thus reduces from at most $(3^3 - 1) \cdot 3^6 \cdot (3^8)^3 = 5\,353\,169\,434\,460\,874 \lesssim 3^{3+6+24} = 3^{33}$ loop iterations to only $\leq 23 \cdot 3^6 \cdot 162^3 = 71\,285\,369\,976$ intent computations, the true number being even further below at $1\,196\,142\,471$.

Once $\mathbb{K}'$ is known, we can employ standard formal concept analysis methods for further simplification, such as object reduction, yielding the context $\mathbb{K}''$ with only 51 objects (see Table 1), or full reduction, leading to the standard context of $\mathbb{K}_1$, which only differs from $\mathbb{K}''$ by deleting the superfluous attribute id_3 and by merging the mutually inverse cyclic permutations $s = (0\,1\,2)$ and $s^{-1} = (2\,1\,0)$ into a single attribute. The standard context ([5], [25, Section 7.4]) is therefore not explicitly shown. We have performed these tasks as well as the computation of all concepts with the help of readily available implementations such as [7, 14].

Remark 16. In order to present the context $\mathbb{K}''$ (see Table 1), we introduce some notation for its objects and attributes. The objects are certain binary and ternary functions on $\{0, 1, 2\}$ that we represent by their value tuples $(a_i)_{0 \leq i < 3^n}$ where $n \in \{2, 3\}$ is the arity of the function. The element $a_i \in \{0, 1, 2\}$ represents the function value $a_i = h(x, y)$ for $i = x + 3y$ and a binary function h, while $a_i = f(x, y, z)$ for $i = x + 3y + 9z$ and a ternary function f, $x, y, z \in \{0, 1, 2\}$.

$$
\begin{array}{llll}
h_0 := 000000000 & h_{11} := 000000222 & h_{17} := 000010002 & h_{25} := 000010222 \\
h_{27} := 000011012 & h_{29} := 000011022 & h_{33} := 000012022 & h_{34} := 000012222 \\
h_{44} := 000110222 & h_{45} := 000111000 & h_{46} := 000111002 & h_{47} := 000111012 \\
h_{51} := 000111111 & h_{52} := 000111112 & h_{54} := 000111122 & h_{55} := 000111202 \\
h_{59} := 000112212 & h_{61} := 000121222 & h_{65} := 000211122 & h_{67} := 000211222 \\
h_{69} := 000212222 & h_{71} := 001110222 & h_{73} := 001111222 & h_{74} := 001211202 \\
h_{76} := 002012222 & h_{78} := 002111022 & h_{80} := 002112012 & h_{88} := 011111112 \\
h_{91} := 011111222 & h_{95} := 012210012 & h_{98} := 012212012 & h_{99} := 012212212 \\
h_{101} := 021210102 & h_{102} := 022212222 & h_{118} := 100100100 & h_{122} := 100111222 \\
h_{123} := 100121220 & h_{129} := 101101101 & h_{131} := 101121121 & h_{133} := 111111111 \\
h_{142} := 200100200 & h_{145} := 200111222 & h_{150} := 220220220 & h_{154} := 220221221 \\
h_{155} := 222222222 & & &
\end{array}
$$

$$
\begin{array}{ll}
f_{167} := 0000100020101110120020122222 & f_{168} := 000011012011111112012112222 \\
f_{169} := 000012022012111212022212222 & f_{171} := 001100200100111121200221222 \\
f_{174} := 000100220101111221220221222 & f_{176} := 000101200101111121220121222
\end{array}
$$

The attributes of $\mathbb{K}''$ are simply all unary functions in $\mathcal{O}_3^{(1)}$. There is a bijection $\varphi \colon \mathcal{O}_3^{(1)} \to \{0, \ldots, 26\}$, mapping each $s \in \mathcal{O}_3^{(1)}$ to $\varphi(s) := 9s(0) + 3s(1) + s(2)$. For $0 \leq j < 3^3 = 27$ we denote $\varphi^{-1}(j) =: s_j$, e.g., $c_0 = s_0$, $id_3 = s_5$ and $c_2 = s_{26}$.

Table 1. Witness-complete row-reduced subcontext $\mathbb{K}''$ of $\mathbb{K}_1$

	s_0	s_1	s_2	s_3	s_4	s_5	s_6	s_7	s_8	s_9	s_{10}	s_{11}	s_{12}	s_{13}	s_{14}	s_{15}	s_{16}	s_{17}	s_{18}	s_{19}	s_{20}	s_{21}	s_{22}	s_{23}	s_{24}	s_{25}	s_{26}
h_0	×	×	×	×	×	×	×	×	×																		
h_{11}	×		×	×		×																		×		×	
h_{17}	×	×	×	×		×	×	×						×													×
h_{25}	×		×	×		×								×										×			×
h_{27}	×	×	×		×	×			×					×	×			×									×
h_{29}	×			×	×		×	×						×													×
h_{33}	×		×	×	×	×		×						×									×	×			×
h_{34}	×		×		×									×								×		×			×
h_{44}	×	×	×		×							×		×											×		×
h_{45}	×		×	×	×					×				×													×
h_{46}	×		×	×	×					×				×													
h_{47}	×		×		×							×		×	×												×
h_{51}	×			×	×				×					×	×												
h_{52}	×			×	×				×					×	×												×
h_{54}	×			×	×				×					×				×				×					×
h_{55}	×		×		×	×				×				×						×							×
h_{59}	×			×	×		×	×	×					×					×								×
h_{61}	×				×			×						×					×					×			×
h_{65}	×				×		×							×				×				×					×
h_{67}	×				×			×						×				×	×			×					×
h_{69}	×				×			×						×				×						×			×
h_{71}	×	×			×							×	×	×													×
h_{73}	×	×			×							×	×	×											×	×	×
h_{74}	×				×									×	×				×								×
h_{76}	×		×	×	×								×	×	×					×			×				×
h_{78}	×		×		×						×			×			×				×		×				×
h_{80}	×	×			×							×		×	×										×	×	×
h_{88}	×			×	×							×	×	×								×	×				×
h_{91}	×			×	×									×	×										×	×	×
h_{95}	×			×	×									×							×	×					×
h_{98}	×			×	×									×			×				×		×				×
h_{99}	×			×		×								×			×					×					×
h_{101}	×			×		×					×			×	×				×		×						×
h_{102}	×			×		×					×			×			×				×		×				×
h_{118}			×	×			×																				
h_{122}			×	×										×	×										×	×	
h_{123}				×											×				×								
h_{129}		×		×						×																	
h_{131}				×													×					×					
h_{133}			×	×	×							×	×	×							×	×	×				
h_{142}				×		×												×									
h_{145}				×		×								×			×					×					×
h_{150}		×		×																			×				
h_{154}				×										×											×		
h_{155}		×		×		×			×					×			×			×			×				×
f_{167}	×	×	×	×		×	×	×		×			×	×	×		×				×			×	×	×	×
f_{168}	×	×	×		×	×			×	×			×	×	×			×	×			×	×		×	×	×
f_{169}	×		×	×	×	×		×	×	×	×		×				×	×	×		×		×	×			×
f_{171}	×			×	×		×	×	×					×				×	×			×					×
f_{174}	×	×	×		×							×	×	×	×										×	×	×
f_{176}	×			×		×	×			×				×			×				×	×		×			×

Table 2. List of 48 conjugacy types of centralising monoids including witnesses

$\{h_0\}^{*(1)} = \{s_0, s_1, s_2, s_3, s_4, s_5, s_6, s_7, s_8\} = s_{15} \circ [M_{9\text{-}4}] \circ s_{15}^{-1}$

$\emptyset^{*(1)} = \mathcal{O}_3^{(1)} = M_{27\text{-}1}$

$\{h_0, h_{17}, f_{167}\}^{*(1)} = \{s_0, s_1, s_2, s_3, s_5, s_6, s_7\} = s_{15} \circ [M_{7\text{-}3}] \circ s_{15}^{-1}$

$\{f_{167}\}^{*(1)} = \{s_0, s_1, s_2, s_3, s_5, s_6, s_7, s_{10}, s_{12}, s_{13}, s_{14}, s_{16}, s_{20}, s_{23}, s_{24}, s_{25}, s_{26}\} = s_{15} \circ [M_{17\text{-}1}] \circ s_{15}^{-1}$

$\{h_{17}, f_{167}\}^{*(1)} = \{s_0, s_1, s_2, s_3, s_5, s_6, s_7, s_{13}, s_{26}\} = s_{15} \circ [M_{9\text{-}3}] \circ s_{15}^{-1}$

$\{h_0, h_{27}, f_{168}\}^{*(1)} = \{s_0, s_1, s_2, s_4, s_5, s_8\} = s_7 \circ [M_{6\text{-}9}] \circ s_7^{-1}$

$\{h_{27}, f_{168}\}^{*(1)} = \{s_0, s_1, s_2, s_4, s_5, s_8, s_{13}, s_{14}, s_{17}, s_{26}\} = s_7 \circ [M_{10\text{-}2}] \circ s_7^{-1}$

$\{h_0, h_{17}, h_{27}, h_{44}, f_{167}, f_{168}, f_{174}\}^{*(1)} = \{s_0, s_1, s_2, s_5\} = s_{21} \circ [M_{4\text{-}6}] \circ s_{21}^{-1}$

$\{f_{174}\}^{*(1)} = \{s_0, s_1, s_2, s_5, s_{11}, s_{12}, s_{13}, s_{14}, s_{24}, s_{25}, s_{26}\} = M_{11\text{-}1}$

$\{f_{167}, f_{168}, f_{174}\}^{*(1)} = \{s_0, s_1, s_2, s_5, s_{12}, s_{13}, s_{14}, s_{24}, s_{25}, s_{26}\} = s_{19} \circ [M_{10\text{-}1}] \circ s_{19}^{-1}$

$\{h_{44}, f_{167}, f_{168}, f_{174}\}^{*(1)} = \{s_0, s_1, s_2, s_5, s_{12}, s_{13}, s_{24}, s_{26}\} = s_{19} \circ [M_{8\text{-}2}] \circ s_{19}^{-1}$

$\{h_{27}, f_{167}, f_{168}, f_{174}\}^{*(1)} = \{s_0, s_1, s_2, s_5, s_{13}, s_{14}, s_{26}\} = s_{21} \circ [M_{7\text{-}4}] \circ s_{21}^{-1}$

$\{h_{17}, h_{27}, h_{44}, f_{167}, f_{168}, f_{174}\}^{*(1)} = \{s_0, s_1, s_2, s_5, s_{13}, s_{26}\} = s_{21} \circ [M_{6\text{-}6}] \circ s_{21}^{-1}$

$\{h_0, h_{17}, h_{27}, h_{44}, h_{71}, h_{73}, f_{167}, f_{168}, f_{174}\}^{*(1)} = \{s_0, s_1, s_5\} = s_{19} \circ [M_{3\text{-}6}] \circ s_{19}^{-1}$

$\{h_{71}, f_{174}\}^{*(1)} = \{s_0, s_1, s_5, s_{11}, s_{12}, s_{13}, s_{26}\} = M_{7\text{-}1}$

$\{h_{44}, h_{71}, h_{73}, f_{167}, f_{168}, f_{174}\}^{*(1)} = \{s_0, s_1, s_5, s_{12}, s_{13}, s_{26}\} = s_{19} \circ [M_{6\text{-}7}] \circ s_{19}^{-1}$

$\{h_{27}, h_{73}, f_{167}, f_{168}, f_{174}\}^{*(1)} = \{s_0, s_1, s_5, s_{13}, s_{14}, s_{26}\} = s_{21} \circ [M_{6\text{-}4}] \circ s_{21}^{-1}$

$\{h_{17}, h_{27}, h_{44}, h_{71}, h_{73}, f_{167}, f_{168}, f_{174}\}^{*(1)} = \{s_0, s_1, s_5, s_{13}, s_{26}\} = s_{19} \circ [M_{5\text{-}4}] \circ s_{19}^{-1}$

$\{h_0, h_{11}, h_{17}, h_{25}, h_{45}, h_{46}, h_{76}, f_{167}\}^{*(1)} = \{s_0, s_2, s_3, s_5\} = s_{15} \circ [M_{4\text{-}5}] \circ s_{15}^{-1}$

$\{h_{45}, h_{46}, f_{167}\}^{*(1)} = \{s_0, s_2, s_3, s_5, s_{10}, s_{13}\} = s_{15} \circ [M_{6\text{-}8}] \circ s_{15}^{-1}$

$\{h_{46}, f_{167}\}^{*(1)} = \{s_0, s_2, s_3, s_5, s_{10}, s_{13}, s_{26}\} = s_{15} \circ [M_{7\text{-}5}] \circ s_{15}^{-1}$

$\{h_{17}, h_{25}, h_{45}, h_{46}, h_{76}, f_{167}\}^{*(1)} = \{s_0, s_2, s_3, s_5, s_{13}\} = s_{15} \circ [M_{5\text{-}6}] \circ s_{15}^{-1}$

$\{h_{17}, h_{25}, h_{46}, h_{76}, f_{167}\}^{*(1)} = \{s_0, s_2, s_3, s_5, s_{13}, s_{26}\} = s_{15} \circ [M_{6\text{-}5}] \circ s_{15}^{-1}$

$\{h_0, h_{11}, h_{17}, h_{25}, h_{27}, h_{44}, h_{45}, h_{46}, h_{47}, h_{76}, h_{80}, f_{167}, f_{168}, f_{174}\}^{*(1)} = \{s_0, s_2, s_5\} = s_{19} \circ [M_{3\text{-}5}] \circ s_{19}^{-1}$

$\{h_{80}, f_{174}\}^{*(1)} = \{s_0, s_2, s_5, s_{11}, s_{13}, s_{14}, s_{24}, s_{25}, s_{26}\} = M_{9\text{-}2}$

$\{h_{47}, h_{80}, f_{174}\}^{*(1)} = \{s_0, s_2, s_5, s_{11}, s_{13}, s_{14}, s_{26}\} = M_{7\text{-}2}$

$\{h_{17}, h_{25}, h_{27}, h_{44}, h_{45}, h_{46}, h_{47}, h_{76}, h_{80}, f_{167}, f_{168}, f_{174}\}^{*(1)} = \{s_0, s_2, s_5, s_{13}\} = s_{19} \circ [M_{4\text{-}3}] \circ s_{19}^{-1}$

$\{h_{80}, f_{167}, f_{168}, f_{174}\}^{*(1)} = \{s_0, s_2, s_5, s_{13}, s_{14}, s_{24}, s_{25}, s_{26}\} = s_{19} \circ [M_{8\text{-}1}] \circ s_{19}^{-1}$

$\{h_{27}, h_{47}, h_{76}, h_{80}, f_{167}, f_{168}, f_{174}\}^{*(1)} = \{s_0, s_2, s_5, s_{13}, s_{14}, s_{26}\} = s_{19} \circ [M_{6\text{-}3}] \circ s_{19}^{-1}$

$\{h_{25}, h_{44}, h_{80}, f_{167}, f_{168}, f_{174}\}^{*(1)} = \{s_0, s_2, s_5, s_{13}, s_{24}, s_{26}\} = s_{19} \circ [M_{6\text{-}2}] \circ s_{19}^{-1}$

$\{h_{17}, h_{25}, h_{27}, h_{44}, h_{46}, h_{47}, h_{76}, h_{80}, f_{167}, f_{168}, f_{174}\}^{*(1)} = \{s_0, s_2, s_5, s_{13}, s_{26}\} = s_{19} \circ [M_{5\text{-}3}] \circ s_{19}^{-1}$

$\{h_{11}, h_{25}, h_{44}, h_{80}, f_{167}, f_{168}, f_{174}\}^{*(1)} = \{s_0, s_2, s_5, s_{24}, s_{26}\} = s_{19} \circ [M_{5\text{-}5}] \circ s_{19}^{-1}$

$\{h_{11}, h_{17}, h_{25}, h_{27}, h_{44}, h_{46}, h_{47}, h_{76}, h_{80}, f_{167}, f_{168}, f_{174}\}^{*(1)} = \{s_0, s_2, s_5, s_{26}\} = s_{19} \circ [M_{4\text{-}2}] \circ s_{19}^{-1}$

$\{h_0, h_{27}, h_{29}, h_{33}, h_{51}, h_{52}, h_{54}, h_{59}, h_{88}, h_{91}, f_{168}, f_{169}, f_{171}\}^{*(1)} = \{s_0, s_4, s_5\} = M_{3\text{-}4}$

$\{h_0, h_{29}, h_{59}, f_{171}\}^{*(1)} = \{s_0, s_4, s_5, s_7, s_8\} = M_{5\text{-}2}$

$\{h_0, h_{27}, h_{29}, h_{33}, h_{59}, f_{168}, f_{169}, f_{171}\}^{*(1)} = \{s_0, s_4, s_5, s_8\} = M_{4\text{-}4}$

$\{h_0, h_{11}, h_{17}, h_{25}, h_{27}, h_{29}, h_{33}, h_{34}, h_{44}, h_{45}, h_{46}, h_{47}, h_{51}, h_{52}, h_{54}, h_{55}, h_{59}, h_{61}, h_{65}, h_{67}, h_{69}, h_{71}, h_{73}, h_{74},$
$\quad h_{76}, h_{78}, h_{80}, h_{88}, h_{91}, h_{95}, h_{98}, h_{99}, h_{101}, h_{102}, f_{167}, f_{168}, f_{169}, f_{171}, f_{174}, f_{176}\}^{*(1)} = \{s_0, s_5\} = M_{2\text{-}1}$

$\{h_0, h_{17}, h_{29}, h_{59}, h_{65}, h_{101}, f_{167}, f_{171}\}^{*(1)} = \{s_0, s_5, s_7\} = s_{15} \circ [M_{3\text{-}2}] \circ s_{15}^{-1}$

$\{h_{101}\}^{*(1)} = \{s_0, s_5, s_7, s_{11}, s_{13}, s_{15}, s_{19}, s_{21}, s_{26}\} = M_{9\text{-}1}$

$\{h_{17}, h_{29}, h_{59}, h_{65}, h_{101}, f_{167}, f_{171}\}^{*(1)} = \{s_0, s_5, s_7, s_{13}, s_{26}\} = s_{15} \circ [M_{5\text{-}1}] \circ s_{15}^{-1}$

$\{h_{17}, h_{25}, h_{27}, h_{29}, h_{33}, h_{34}, h_{44}, h_{45}, h_{46}, h_{47}, h_{51}, h_{52}, h_{54}, h_{55}, h_{59}, h_{65}, h_{67}, h_{69}, h_{71}, h_{73}, h_{74}, h_{76}, h_{78}, h_{80},$
$\quad h_{88}, h_{91}, h_{95}, h_{98}, h_{99}, h_{101}, h_{102}, f_{167}, f_{168}, f_{169}, f_{171}, f_{174}, f_{176}\}^{*(1)} = \{s_0, s_5, s_{13}\} = M_{3\text{-}3}$

$\{h_{74}, h_{101}\}^{*(1)} = \{s_0, s_5, s_{13}, s_{15}, s_{19}, s_{26}\} = M_{6\text{-}1}$

$\{h_{17}, h_{25}, h_{27}, h_{29}, h_{33}, h_{34}, h_{44}, h_{46}, h_{47}, h_{52}, h_{54}, h_{55}, h_{59}, h_{65}, h_{67}, h_{69}, h_{71}, h_{73}, h_{74}, h_{76}, h_{78}, h_{80}, h_{88}, h_{91},$
$\quad h_{95}, h_{98}, h_{99}, h_{101}, h_{102}, f_{167}, f_{168}, f_{169}, f_{171}, f_{174}, f_{176}\}^{*(1)} = \{s_0, s_5, s_{13}, s_{26}\} = M_{4\text{-}1}$

$\{h_0, h_{11}, h_{17}, h_{25}, h_{27}, h_{44}, h_{45}, h_{46}, h_{47}, h_{76}, h_{80}, h_{150}, h_{155}, f_{167}, f_{168}, f_{174}\}^{*(1)} = \{s_2, s_5\} = s_{19} \circ [M_{2\text{-}2}] \circ s_{19}^{-1}$

$\{h_0, h_{27}, h_{155}, f_{168}\}^{*(1)} = \{s_2, s_5, s_8\} = s_7 \circ [M_{3\text{-}8}] \circ s_7^{-1}$

$\{h_{11}, h_{25}, h_{44}, h_{80}, h_{150}, f_{167}, f_{168}, f_{174}\}^{*(1)} = \{s_2, s_5, s_{24}\} = s_{19} \circ [M_{3\text{-}7}] \circ s_{19}^{-1}$

$\{h_0, h_{11}, h_{17}, h_{25}, h_{27}, h_{29}, h_{33}, h_{34}, h_{44}, h_{45}, h_{46}, h_{47}, h_{51}, h_{52}, h_{54}, h_{55}, h_{59}, h_{61}, h_{65}, h_{67}, h_{69}, h_{71}, h_{73}, h_{74},$
$\quad h_{76}, h_{78}, h_{80}, h_{88}, h_{91}, h_{95}, h_{98}, h_{99}, h_{101}, h_{102}, h_{118}, h_{122}, h_{123}, h_{129}, h_{131}, h_{133}, h_{142}, h_{145}, h_{150}, h_{154}, h_{155},$
$\quad f_{167}, f_{168}, f_{169}, f_{171}, f_{174}, f_{176}\}^{*(1)} = \{s_5\} = M_{1\text{-}1}$

$\{h_{74}, h_{101}, h_{123}\}^{*(1)} = \{s_5, s_{15}, s_{19}\} = M_{3\text{-}1}$

Relying on $\mathbb{K}''$ and [7], it is now straightforward to confirm the following [5]:

Theorem 17. (cf. [19, Proposition 3.5]). *The number of centralising monoids on $\{0,1,2\}$ (intents of $\mathbb{K}'$, $\mathbb{K}''$) is 192; they fall into 48 classes up to conjugacy.*

Proof. The centralising monoids on $\{0,1,2\}$ are the intents of $\mathbb{K}_1$. As the set W computed above is witness-complete, the finite context $\mathbb{K}'$ has the same intents as $\mathbb{K}_1$, and likewise for the object-reduced subcontext $\mathbb{K}''$. Using [7] on $\mathbb{K}''$, we get a list of 192 concepts [25, 7.5]; their intents are the centralising monoids, their extents provide witnesses that are not available from [19,21]. We iterate over this list and compute all (up to 5) non-identical conjugates of each monoid, directly removing them from the list [5]. Upon termination, 48 distinct intents remain.   □

Table 2 presents the 48 representatives (cf. [5]) of all centralising monoids on the set $\{0,1,2\}$ up to conjugacy, together with a witness for each one and a hint from which row of [19, Table 3] the monoid can be obtained by conjugation with which permutation $s \in \mathrm{Sym}(3) = \{s_5, s_7, s_{11}, s_{15}, s_{19}, s_{21}\}$.

Acknowledgements. The authors are grateful to the anonymous referees for their comments and the careful reading of the manuscript.

References

1. Behrisch, M.: All centralising monoids with binary operations as witnesses on the set $\{0,1,2,3\}$. Zenodo (2021). https://doi.org/10.5281/zenodo.5428986
2. Behrisch, M.: Centralising monoids with conservative majority operations as witnesses. In: Proceedings 51st ISMVL 2021, Nur-Sultan, Kazakhstan, 25–27 May 2021, pp. 56–61. IEEE, Los Alamitos (2021). https://doi.org/10.1109/ISMVL51352.2021.00019
3. Behrisch, M.: All centralising monoids with majority witnesses on a four-element set. J. Mult.-Valued Logic Soft Comput. **38**(1–2), 23–56 (2022), https://www.oldcitypublishing.com/journals/mvlsc-home/mvlsc-issue-contents/mvlsc-volume-38-number-1-2-2022/mvlsc-38-1-2-p-23-56/
4. Behrisch, M., Couceiro, M., Kearnes, K.A., Lehtonen, E., Szendrei, Á.: Commuting polynomial operations of distributive lattices. Order **29**(2), 245–269 (2012). https://doi.org/10.1007/s11083-011-9231-3
5. Behrisch, M., Renkin, L.: All centralising monoids on the set $\{0,1,2\}$, including their witnesses. Zenodo (2023). https://doi.org/10.5281/zenodo.7641814
6. Behrisch, M., Vargas-García, E.: Centralising monoids with low-arity witnesses on a four-element set. Symmetry **13**(8), 1471:1–40 (2021). https://doi.org/10.3390/sym13081471
7. Borchmann, D., Hanika, T.: ConExp-clj (2023). https://github.com/tomhanika/conexp-clj. Accessed 20 Apr 2023
8. Burris, S., Willard, R.: Finitely many primitive positive clones. Proc. Am. Math. Soc. **101**(3), 427–430 (1987). https://doi.org/10.2307/2046382

9. Couceiro, M., Lehtonen, E.: Explicit descriptions of bisymmetric Sugeno integrals. In: Hüllermeier, E., Kruse, R., Hoffmann, F. (eds.) IPMU 2010. LNCS (LNAI), vol. 6178, pp. 494–501. Springer, Heidelberg (2010). https://doi.org/10.1007/978-3-642-14049-5_51

10. Couceiro, M., Lehtonen, E.: Self-commuting lattice polynomial functions on chains. Aequationes Math. **81**(3), 263–278 (2011). https://doi.org/10.1007/s00010-010-0058-6

11. Goldstern, M., Machida, H., Rosenberg, I.G.: Some classes of centralizing monoids on a three-element set. In: Proceedings 45th ISMVL 2015, Waterloo, ON, Canada, 18–20 May 2015, pp. 205–210. IEEE, Los Alamitos (2015). https://doi.org/10.1109/ISMVL.2015.29

12. Grabisch, M., Marichal, J.-L., Mesiar, R., Pap, E.: Aggregation functions: construction methods, conjunctive, disjunctive and mixed classes. Inf. Sci. **181**(1), 23–43 (2011). https://doi.org/10.1016/j.ins.2010.08.040

13. Grabisch, M., Marichal, J.-L., Mesiar, R., Pap, E.: Aggregation functions: means. Inf. Sci. **181**(1), 1–22 (2011). https://doi.org/10.1016/j.ins.2010.08.043

14. Hanika, T., Hirth, J.: Conexp-Clj - A research tool for FCA. In: Cristea, D., Le Ber, F., Missaoui, R., Kwuida, L., Sertkaya, B. (eds.) Supplementary Proceedings ICFCA 2019, Frankfurt, Germany, 25–28 June 2019. CEUR Workshop Proceedings, vol. 2378, pp. 70–75. CEUR-WS.org (2019). https://ceur-ws.org/Vol-2378/shortAT8.pdf

15. Ježek, J., Kepka, T.: Equational theories of medial groupoids. Algebra Univers. **17**(2), 174–190 (1983). https://doi.org/10.1007/BF01194527

16. Machida, H., Rosenberg, I.G.: On endoprimal monoids in clone theory. In: Proceedings 39th ISMVL 2009, Naha, Okinawa, Japan, 21–23 May 2009, pp. 167–172. IEEE, Los Alamitos (2009). https://doi.org/10.1109/ISMVL.2009.53

17. Machida, H., Rosenberg, I.G.: Endoprimal monoids and witness lemma in clone theory. In: Proceedings 40th ISMVL 2010, Barcelona, Spain, 26–28 May 2010, pp. 195–200. IEEE, Los Alamitos (2010). https://doi.org/10.1109/ISMVL.2010.44

18. Machida, H., Rosenberg, I.G.: Maximal centralizing monoids and their relation to minimal clones. In: Proceedings 41st ISMVL 2011, Tuusula, Finland 23–25 May 2011, pp. 153–159. IEEE, Los Alamitos (2011). https://doi.org/10.1109/ISMVL.2011.36

19. Machida, H., Rosenberg, I.G.: Centralizing monoids on a three-element set. In: Miller, D.M., Gaudet, V.C. (eds.) Proceedings 42nd ISMVL 2012, Victoria, BC, Canada, 14–16 May 2012, pp. 274–280. IEEE, Los Alamitos (2012). https://doi.org/10.1109/ISMVL.2012.50

20. Machida, H., Rosenberg, I.G.: Some centralizing monoids on a three-element set. J. Mult.-Valued Logic Soft Comput. **18**(2), 211–221 (2012). https://www.oldcitypublishing.com/journals/mvlsc-home/mvlsc-issue-contents/mvlsc-volume-18-number-2-2012/mvlsc-18-2-p-211-221/

21. Machida, H., Rosenberg, I.G.: Report on centralizing monoids on a three-element set. In: Machida, H. (ed.) Clone Theory and Discrete Mathematics, June 13–15 2005, Algebra and Logic Related to Computer Science, 22–23 October 2009. RIMS Kôkyûroku, vol. 1846, pp. 53–65. RIMS, Kyoto University, Kyoto, Japan (2013). https://www.kurims.kyoto-u.ac.jp/~kyodo/kokyuroku/contents/1846.html

22. Machida, H., Rosenberg, I.G.: Centralizing monoids and the arity of witnesses. In: Proceedings 47th ISMVL 2017, Novi Sad, Serbia, 22–24 May 2017, pp. 236–241. IEEE, Los Alamitos (2017). https://doi.org/10.1109/ISMVL.2017.34

23. Murdoch, D.C.: Structure of abelian quasi-groups. Trans. Am. Math. Soc. **49**(3), 392–409 (1941). https://doi.org/10.2307/1989940

24. Pöschel, R., Kalužnin, L.A.: Funktionen- und Relationenalgebren. Deutscher Verlag der Wissenschaften, Berlin (1979). https://doi.org/10.1007/978-3-0348-5547-1

25. Renkin, L.: Centralizing Monoids on a Three-Element Set. Bachelor's thesis, Technische Universität Wien (2022). https://doi.org/10.5281/zenodo.7653085

Description Quivers for Compact Representation of Concept Lattices and Ensembles of Decision Trees

Egor Dudyrev[1,2]([✉])[iD], Sergei O. Kuznetsov[1][iD], and Amedeo Napoli[2][iD]

[1] HSE University, Moscow, Russia
[2] Université de Lorraine, CNRS, Inria, LORIA, 54000 Nancy, France
`egor.dudyrev@yandex.ru`

Abstract. In this paper we introduce and study *description quivers* as compact representations of concept lattices and respective ensembles of decision trees. Formally, description quivers are directed multigraphs where vertices represent concept intents and (multiple) edges represent generators of intents. We study some properties of description quivers and shed light on their use for describing state-of-the-art symbolic machine learning models based on decision trees. We also argue that a concept lattice can be considered as a cornerstone in constructing an efficient machine learning model. We show that the proposed description quivers allow us to fuse decision trees just as we can sum linear regressions, while proposing a way to select the most important rules in decision models, just as we can select the most important coefficients in regressions.

Keywords: Formal Concept Analysis · Supervised Machine Learning · Explainable Artificial Intelligence

1 Introduction

Intents and their generators are important tools of Formal Concept Analysis [9]. This paper introduces generators of difference between comparable intents and proposes some ways of using them, such as simplifying the lattice visualization and summation of decision trees.

Formal Concept Analysis (FCA) is a mathematically well-founded theory aimed at data analysis. One of its tools for analysing data consists in providing computable representations of a dataset [6]. These are, e.g., intents (the maximal subsets of attributes describing a specific subset of objects), and their generators (subsets of intents that describe the same subset of objects as intents). This paper generalises the notion of generators to difference generators that describe the same subset of objects as the difference of two comparable intents. Here we study difference generators and provide some of their use-cases.

First, the difference generators help data analysis by presenting more concise line diagrams of concept lattices given by intents. Intents, being maximal

© The Author(s), under exclusive license to Springer Nature Switzerland AG 2023
D. Dürrschnabel and D. López Rodríguez (Eds.): ICFCA 2023, LNAI 13934, pp. 127–142, 2023.
https://doi.org/10.1007/978-3-031-35949-1_9

subsets of attributes corresponding to a set of objects often make diagrams too overloaded with text. The well-known solution to this problem is to show only the "new" (w.r.t. to predecessors intents) attributes of intents, but not any of its smaller intents. Difference generators allow to present only the key new attributes of an intent. Thus, they require less text on a line diagram of a lattice of intents and help to highlight the lattice structure.

Difference generators also make it possible to introduce description quivers as a graph theory-based view of a lattice of intents. A description quiver is a directed multigraph, also known as a "quiver", with intents as nodes and difference generators between intents as edges. This graph-theoretic definition proposes a useful interface between FCA and graph theory. Moreover, due to a versatile choice of its edges, description quivers can both copy the "behaviour" of a lattice of intents, or refer only to a subset of attributes. Therefore, description quivers can be studied as mathematical objects on their own. This paper introduces the first results of such a study.

Finally, description quivers can be easily applied in a supervised machine learning scenario, resulting in decision quiver model. This allows us to describe decision trees [5] and their ensembles (e.g. random forest [4]) in terms of FCA. The connections between FCA and decision trees were extensively studied in [1–3,7,10,11], and [12].

Accordingly, this paper presents a new simpler notation to describe connections between FCA and decision trees. The notation also presents a simple way to combine an ensemble of loosely connected decision trees into one interconnected model. This paper continues the work started in [7] and [8]. It merges these two works and proposes a better mathematical language for describing relations between FCA and decision trees.

The paper is organised as follows. Section 2.1 recalls basic definitions of Formal Concept Analysis (FCA). Then Sect. 2.2 introduces difference generators and how to use them for providing more concise visualizations of lattices of intents. Section 3 introduces and studies description quiver: a directed multidigraph with intents as nodes and difference generators as edges. Section 4 introduces decision quivers that allow summation of decision trees. Experimental results presented in Sect. 5 and Sect. 6 concludes the paper.

2   Theoretical Background

2.1   Formal Concept Analysis

This section recalls basic definitions of Formal Concept Analysis to facilitate their generalizations in the following sections.

As usual, data are given in the form of a **formal context** $K = (G, M, I)$, where G is the set of objects, M is the set of (binary) attributes, and I is a relation between objects and attributes: $I \subseteq G \times M$.

For a given subset of attributes $X \subseteq M$ and an attribute $m \in M$, we often want to consider the cases when attribute m belongs to $X : m \in X$, when attribute m does not belong to $X : m \notin X$, also represented as $\overline{m} \in X$, and when the presence of attribute m is irrelevant: $m \notin X$ and $\overline{m} \notin X$. Therefore, we define a dichotomised set of attributes $M^{\pm}$ as the union of attributes M and their negations:

$$M^{\pm} = \{m, \overline{m} \mid m \in M\}. \tag{1}$$

Standard **prime operators** are defined as follows: A' gives the maximal subset of attributes (i.e. a **description**) shared by all objects from $A \subseteq G$; analogously, B' gives the maximal subset of objects (that we call an **extent**) shared by all attributes from $B \subseteq M$.

$$A' = \{m \in M^{\pm} \mid \forall g \in A : gIm\}, \quad A \subseteq G \tag{2}$$

$$B' = \{g \in G \mid \forall m \in B : gIm\}, \quad B \subseteq M^{\pm} \tag{3}$$

Formal Concept Analysis studies formal concepts and the partial order (lattice) over them. A formal concept if a pair (A, B) of subsets of objects A and attributes B, such that A is the extent of $B : A = B'$, and B is the **intent** (i.e. maximal description) of $A : B = A'$. To make the notation of this paper concise we will ignore the extents and concentrate only on intents. It is well-known that the set of all intents forms a lattice dual to the lattice of extents:

$$\mathbb{L} = \{B \subseteq M^{\pm} \mid B'' = B\} \tag{4}$$

Each intent $B \subseteq M^{\pm} : B'' = B$ corresponds to many subsets of attributes $D \subseteq M^{\pm}$ with the same closure $B : D'' = B$. Such subset of attributes D is called a **generator** of B. The set of all generators of B gives the **equivalence class** of $B : [B] = \{D \subseteq M^{\pm} \mid D'' = B\}$. A subset of attributes $D \subseteq M^{\pm}$ is called a **key** (or a **minimal generator**) if it is the smallest subset of attributes with closure B: i.e. $\forall E \subset D : E'' \neq D'' = B$. Note that for each intent $B \subseteq M^{\pm}$ there can be multiple (even exponentially many) incomparable keys.

Table 1 presents a formal context that is going to be a running example in the paper. The purpose for the target column Y (so as the reason for using numbers 0 and 1 instead of crosses) will be covered in Sect. 4. This formal context is inspired by the classic "Live in water" formal context [13].

2.2   Difference Generators

The previous subsection describes our motivation to study the keys of set differences between comparable intents. This subsection defines these differences mathematically.

Definition 1. *Subset of attributes D is called a **difference generator** between intents B_2, B_1, such that $B_1 \subseteq B_2$, if its closure is equal to the closed difference between intents B_2 and B_1:*

$$D \subseteq M^{\pm} : D'' = (B_2 \setminus B_1)'', \quad \text{where } B_1, B_2 \in \mathbb{L}, B_1 \subseteq B_2. \tag{5}$$

Table 1. Running example of a formal context with additional target column Y and test objects

		Attributes M				Target Y
		l	w	c—	h	
Objects G	dog	×		×	×	1
	corn	×				0
	bream		×	×	×	0
	egg					0
Test objects	reed	×	×			0
	sea snake	×	×	×		0
Attr. names		lives on land	lives in water	can move	has limbs	breast feeds

A trivial example of a difference generator between intents B_2 and B_1 is the difference $B_2 \setminus B_1$ itself. In fact, we are not interested in any difference generator larger than the difference.

Proposition 1. *Subset D of the difference between two comparable intents B_2 and B_1 is a difference generator between these intents if and only if the union of its closure with B_1 is the closure of B_2:*

$$\forall B_1, B_2 \in \mathbb{L}, B_1 \subseteq B_2, D \subseteq B_2 \setminus B_1 : D'' = (B_2 \setminus B_1)'' \iff B_1 \cup D'' = B_2 \quad (6)$$

Proof. Let us derive prove the proposition in two directions:

- From left to right:
  If $D'' = (B_2 \setminus B_1)''$ then $(B_2 \setminus B_1) \subseteq (B_2 \setminus B_1)'' = D''$ and $D'' \subseteq B_2$
  therefore $B_1 \cup D'' = B_2$;
- From right to left:
  First, if $B_1 \cup D'' = B_2$ then $(B_2 \setminus B_1) \subseteq D''$;
  second, if $D \subseteq B_2 \setminus B_1$ then $D'' \subseteq (B_2 \setminus B_1)''$;
  therefore $D'' = (B_2 \setminus B_1)''$.

Note that any generator $D \subseteq M^{\pm}$ of an intent $B \subseteq M^{\pm} : D'' = B$ can be represented as a difference generator between B and the minimal intent $\emptyset''$: $D'' = B \iff D'' = (B \setminus \emptyset'')''$.

Intent differences often used in FCA to shorten the labels of nodes in a line diagram of lattice of intents. That is, instead of presenting the full intent B in a lattice L, one only shows the "new" attributes of the intent, that are not included in smaller intents of the lattice: $B \setminus (\bigcup_{\tilde{B} \in L, \tilde{B} \subset B} \tilde{B})$. However, such set differences can also contain many attributes. So keys, i.e., minimal generators of the differences, can be used.

Figure 1 presents these three ways to label the nodes on a line diagram. The left plot labels the nodes with the full intents, the middle plot only shows new attributes, contained in the intent, and the right plot gives a key of the set of "new" attributes. The lattice in the figure is based on ten intents having the biggest values of stability lower bound for the Zoo dataset (plus the smallest and the biggest intents). It can be seen that the right plot shows the minimal set of attributes, thus presenting only the most important ones. A qualitative comparison of label lengths is presented in section Experiments.

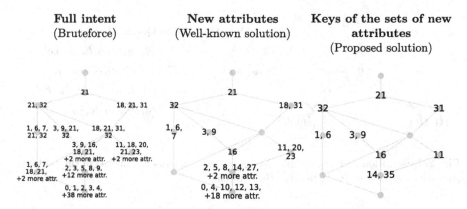

Fig. 1. Three ways to denote intents in a lattice. Given lattice represents ten most stable intents in Zoo dataset (plus top and bottom intents). Numbers stand for indices of binary attributes. The left subfigure visualizes indices of all attributes from intents that makes the diagram almost incomprehensible. In contrast, the right subfigure shows only a few attribute indices that makes it easier to read the diagram and also allows to increase the font size.

3 Description Quivers for Unsupervised Setting

In the previous section introduced the basic definitions of FCA, the task we solve, and the difference keys aimed at connecting the FCA terms of intents and their keys. This section is devoted to describing description quiver: a graph-like model that merges intents and difference keys. Description quiver is designed to balance the length of intents with the redundancy of equivalent keys.

3.1 Description Quiver

Definition 2. Description quiver (L, E) *is a pair of subset of intents L and a subset of difference generators E between intents L. If the directed multigraph (L, E) is weakly-connected, it is called a* description quiver:

$$(L, E) \text{ is a description quiver, where}$$

- $L \subseteq \mathbb{L} : \emptyset'' \in L$
- $E = \{(B_1, B_2, D) \mid B_1 \cup D'' = B_2,$
$$B_1, B_2 \in L, B_1 \subset B_2, D \subseteq B_2 \setminus B_1\}$$

(7)

An example of a description quiver for the context from Table 1 is given on Fig. 2. Note that a quiver can contain have multiple edges between the same nodes as there can be multiple difference generators between two intents.

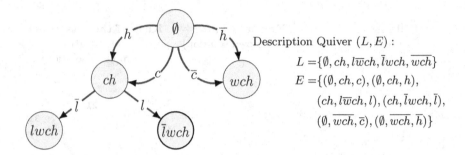

Description Quiver (L, E) :

$$L = \{\emptyset, ch, l\overline{w}ch, \overline{l}wch, \overline{wch}\}$$
$$E = \{(\emptyset, ch, c), (\emptyset, ch, h),$$
$$(ch, l\overline{w}ch, l), (ch, \overline{l}wch, \overline{l}),$$
$$(\emptyset, \overline{wch}, \overline{c}), (\emptyset, \overline{wch}, \overline{h})\}$$

Fig. 2. Description Quiver example. Nodes represent a subset of intents of the contexts. And edges show how to "generate" these intents

3.2 Path in Quiver

As for any graph, we can define traversal procedure for a decision quiver. This subsection introduces paths in quivers adapted to intents and difference generators (i.e., to decision quivers).

First, let us define a path in quiver (L, E).

Definition 3. *Given quiver (L, E), a sequence of k edges $\langle e_i \rangle_{i=1}^{k}, e_i \in E$ is called a **path in the quiver** if each element e_i of the path is a difference generator between intents B_i and B_{i-1}, and the first element e_1 is a difference generator between B_1 and $\emptyset''$:*

$$\langle e_i \rangle_{i=1}^{k} \subseteq E : \forall i = 1, \ldots, k : e_i = (B_{i-1}, B_i, D_i), B_0 = \emptyset'' \tag{8}$$

For example, consider the quiver in Fig. 2. Its five intents and six edges are given on the right of the figure. An example of a path in this quiver would be the tuple of edges $\langle (\emptyset, ch, c), (ch, l\overline{w}ch, l) \rangle$.

Now we can define a set of paths P in quiver (L, E) as follows:

Definition 4. *Given quiver (L, E), **set of paths** P in the quiver is the set of all possible paths in the quiver:*

$$P = \{\langle e_i \rangle_{i=1}^{k} \subseteq E \mid \forall i = 1, \ldots, k : e_i = (B_{i-1}, B_i, D_i), B_0 = \emptyset'', k \in \mathbb{N}\}. \tag{9}$$

For the quiver in Fig. 2, the set of paths $P = \{ \langle (\emptyset, ch, c), (ch, l\overline{w}ch, l) \rangle, \langle (\emptyset, ch, h), (ch, l\overline{w}ch, l) \rangle, \langle (\emptyset, ch, c), (ch, \overline{l}wch, \overline{l}) \rangle, \langle (\emptyset, ch, h), (ch, \overline{l}wch, \overline{l}) \rangle, \langle (\emptyset, \overline{wch}, \overline{c}) \rangle, \langle (\emptyset, \overline{wch}, \overline{h}) \rangle \}$.

Definition 5. *Given quiver* (L, E) *and description* $X \subseteq M$, *the **set of paths*** P_X , ***following description*** X, *is the set of maximal paths in quiver* (L, E) *such that each edge description in a path is a subset of* X:

$$P_{X,any \ k} = \{\langle e_i \rangle_{i=1}^{k} \in P \mid D_i \subseteq X, \forall i \in \mathbb{N} : i \leq k\},$$

$$P_X = \{\langle e_i \rangle_{i=1}^{k} \in P_{X,any \ k} \mid \nexists \langle e_i \rangle_{i=1}^{k+1} \in P_{X,any \ k}\}, \tag{10}$$

where $e_i = (B_{i-1}, B_i, D_i)$.

For the quiver in Fig. 2, the set of paths P_{lwc}, following description lwc would be $P_{lwc} = \{\langle (\emptyset, ch, c), (ch, \overline{lw}ch, l) \rangle, \langle (\emptyset, ch, h), (ch, \overline{lw}ch, l) \rangle, \langle (\emptyset, \overline{wch}, \overline{h}) \rangle\}$.

We are specifically interested in the set of intents L_X we can arrive to via paths P_X, and the set of edges we pass E_X while traversing the quiver:

Definition 6. *Given quiver* (L, E) *and description* $X \subseteq M$, *the subset of the terminal intents* L_X *in paths, following* X, *is called the set of **targets** of* X. *The set of **maximal targets** of* X *is denoted by* $L_{X,\max}$. *The set of edges* E_X *from paths* P_X *is called the set of **arrows** of* X.

$$L_X = \{B_k \mid \langle e_i \rangle_{i=1}^{k} \in P_X\},$$

$$L_{X,\max} = \{B \in L_X \mid \nexists \tilde{B} \in L_X : B \subset \tilde{B}\} \tag{11}$$

$$E_X = \{e_1, \ldots, e_k \mid \langle e_i \rangle_{i=1}^{k} \in P_X\}$$

where $e_i = (B_{i-1}, B_i, D_i)$.

For the quiver in Fig. 2, the targets L_{lwc} of description lwc are $L_{lwc} = \{\overline{lw}ch, \overline{wch}\}$. The targets are incomparable, therefore the set of maximal targets of description lwc is the same: $L_{lwc,\max} = L_{lwc}$. The arrows of description lwc is $E_{lwc} = \{(\emptyset, ch, c), (ch, \overline{lw}ch, l), (\emptyset, \overline{wch}h)\}$.

Definition 7. *Description quiver* (L, E) *is called a **description tree** if for each non-top node of the quiver there is only one node leading to it:*

$$(L, E) \text{ is a tree } \iff (L, E) \text{ is a quiver, and}$$
$$\forall B_2 \in L \setminus \{\emptyset''\} : \left| \{B_1 \mid (B_1, B_2, D) \in E\} \right| = 1 \tag{12}$$

Description tree (L, E) is called **dichotomic** when each node $B \in L$ with edges, going from this node, $\{(B_1, B_2, D) \in E \mid B_1 = B\} \neq \emptyset$ has exactly two such edges and their descriptions are dichotomic attributes $m, \overline{m} \in M^{\pm}$: $\{(B_1, B_2, D) \in E \mid B_1 = B\} = \{(B, B_3, \{m\}), (B, B_4, \{\overline{m}\})\}$, where $B_3, B_4 \in L$.

4   Decision Quivers for Supervised Setting

In this subsection we show the connections between description quivers and decision trees, a basic supervised machine learning model.

In supervised machine learning setup, we are given a context (G, M, I), the set of outcomes Y and the mapping $\tau : G \rightarrow Y$ from (training) objects outcomes. The set of outcomes is usually defined as $Y = \{0, 1\}$ for binary classification task, or as a set of real values $\mathbb{R}$ for regression task. Both tasks aims to finding a function $\varphi : 2^M \rightarrow Y$ that maps each description $X \subseteq M$ to a value from Y. Prediction function φ is chosen w.r.t. prediction quality measures that compare the outcomes of τ and φ on objects G and their descriptions.

In this subsection we study two types of decision quivers: target-based decision quiver $\dot{Q} = (L, E, \dot{\varphi} : L \rightarrow Y)$, and arrow-based decision quiver $\vec{Q} = (L, E, \vec{\varphi} : D \rightarrow Y)$. The small dots above the signs for $\dot{Q}$ and $\dot{\varphi}$ symbolize that target-based quivers assign predictions to the nodes L. Analogously, arrow-based quivers assign predictions to the edges E that is highlighted by arrow above the signs for $\vec{Q}$ and $\vec{\varphi}$.

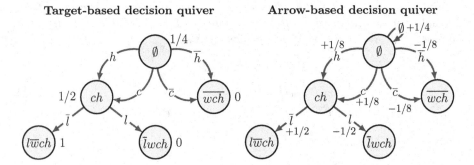

Fig. 3. Two types of Decision Quivers: target-based decision quiver assigns labels (or predictions) to the nodes, and arrow-based decision quiver assigns labels to the edges. Here, the numbers represent the probability of an object being "breast-feeding" (target Y) based on its attributes M.

4.1   Target-based Decision Quiver

Definition 8. *Target-based decision quiver is a triple $(L, E, \dot{\varphi} : L \rightarrow Y)$, where the pair (L, E) makes a description quiver and $\dot{\varphi}$ is a function that maps intents from L to values from Y. We will denote target-based decision quivers by $\dot{Q}$:*

$$\dot{Q} = (L, E, \dot{\varphi} : L \rightarrow Y) \tag{13}$$

To make a prediction $\dot{Q}(X)$ for description $X \subseteq M$ with target-based decision quiver $\dot{Q} = (L, E, \dot{\varphi})$, we average the individual predictions of maximal targets of description X:

$$\dot{Q}(X) = \frac{1}{|L_{X,\max}|} \sum_{B \in L_{X,\max}} \dot{\varphi}(B). \tag{14}$$

Target-based decision quivers make it easy to define a machine learning model in the FCA framework. First, we select a subset of intents L based on some interestingness measure. Second, we define a prediction function $\dot{\varphi}$ for each intent $B \in L$ (e.g. $\dot{\varphi}(B) = \frac{1}{|B'|} \sum_{g \in B'} \tau(g)$). Third, we select a subset of directed generators D to connect intents from L.

In fact, decision tree – a basic machine learning model – can be represented as a target-based decision quiver.

Proposition 2. *Decision tree is a target-based decision quiver* $(L, E, \dot{\varphi})$ *where* (L, E) *forms a description tree.*

4.2   Arrow-Based Decision Quiver

Definition 9. *Arrow-based decision quiver is a triple* $(L, E, \vec{\varphi} : E \to Y)$, *where the pair* (L, E) *is a description quiver and* $\vec{\varphi}$ *is a function that maps directed generators from* D *to values from* Y. *We denote arrow-based decision quivers by* $\vec{Q}$:

$$\vec{Q} = (L, E, \vec{\varphi} : E \to Y) \tag{15}$$

To make a prediction $\vec{Q}(X)$ for description $X \subseteq M$ with arrow-based decision quiver $\vec{Q} = (L, E, \vec{\varphi})$, we sum the predictions $\vec{\varphi}$ of all arrows E_X following description X:

$$\vec{Q}(X) = \sum_{e \in E_X} \vec{\varphi}(e). \tag{16}$$

Often the models are required to make nonzero basic prediction. In target-based decision quivers such predictions are expressed by the prediction of the top intent $\dot{\varphi}(\emptyset'')$. In arrow-based decision quivers this can be represented with the trivial edge $(\emptyset'', \emptyset'', \emptyset)$: $\vec{\varphi}((\emptyset'', \emptyset'', \emptyset))$.

4.3   Summation of Arrow-Based Decision Quivers

Let us define a summation operation on arrow-based decision quivers:

$$\vec{Q}_1 + \vec{Q}_2 = \vec{Q}_\Sigma,$$
$$\text{where } \vec{Q}_1 = (L_1, E_1, \vec{\varphi}_1), \ \vec{Q}_2 = (L_2, E_2, \vec{\varphi}_2) \tag{17}$$
$$\vec{Q}_\Sigma = (L_1 \cup L_2, E_1 \cup E_2, \vec{\varphi}_1 + \vec{\varphi}_2)$$

Arrow-based decision quivers can be multiplied by scalar:

$$(L, E, \vec{\varphi}) \cdot k = (L, E, k \cdot \vec{\varphi}), \quad \text{where } k \in \mathbb{R} \tag{18}$$

Therefore, arrow-based decision quivers allow us to "sum up" many quivers to one. Thus, they form a vector space.

Let us study how predictions of the sum of quivers relate to the predictions of the initial quivers. Consider an example of the averaging of two arrow-based decision quivers on Fig. 4. Figure represents two quivers $\vec{Q}_1 = (L_1, E_1, \vec{\varphi}_1)$, $\vec{Q}_2 = (L_2, E_2, \vec{\varphi}_2)$ and their average quiver $\vec{Q}_3 = (\vec{Q}_1 + \vec{Q}_2)/2 = (L_3, E_3, \vec{\varphi}_3)$. For object *bream* with description *wch* the quivers predictions will be $\vec{Q}_1(wch) = 0, \vec{Q}_2(wch) = 0, \vec{Q}_3(wch) = 0$.

Therefore, the average of two initial quivers predictions $(\vec{Q}_1(wch) + \vec{Q}_2(wch))/2$ equals the prediction of the averaged quiver $\vec{Q}_3(wch)$. This property does not hold for any description $X \subseteq M$. Consider object *sea snake* with description *lwc*, the predictions will be $\vec{Q}_1(lwc) = 1, \vec{Q}_2(lwc) = 0, \vec{Q}_3(lwc) = 3/4$.

In this case, the average of two initial predictions $(\vec{Q}_1(lwc) + \vec{Q}_2(lwc))/2$ is not equal to the prediction of the averaged quiver $\vec{Q}_3(lwc)$. However, the latter lies between the two initial predictions $\vec{Q}_2(lwc) < \vec{Q}_3(lwc) < \vec{Q}_1(lwc)$.

This problem of keeping predictions the same after the averaging requires an extensive study. For now, we can formulate the following proposition:

Proposition 3. *Given two arrow-based decision quivers* $\vec{Q}_1 = (L_1, E_1, \vec{\varphi}_1)$, $\vec{Q}_2 = (L_2, E_2, \vec{\varphi}_2)$ *with only one common intent:* $L_1 \cap L_2 = \{\emptyset''\}$, *the prediction of the sum* $\vec{Q}_\Sigma(X) = (\vec{Q}_1 + \vec{Q}_2)(X)$ *is equal to the sum of predictions* $\vec{Q}_1(X) + \vec{Q}_2(X)$ *for any description* $X \subseteq M$.

Proof. Without loss of generality, assume that both initial quivers have the trivial edge $e_t = (\emptyset'', \emptyset'', \emptyset)$ with labels $\vec{\varphi}_1(e_t), \vec{\varphi}_2(e_t)$ that can be equal to zero.

Let the sum quiver be $\vec{Q}_\Sigma = (L_\Sigma, E_\Sigma, \vec{\varphi}_\Sigma)$. Since $\emptyset''$ is the only intent contained in both quivers, the only edge containing in both quivers would be the trivial edge $E_1 \cap E_2 = \{e_t\}$. Therefore, by the definition of the sum operation, any edge $e \in E_1 \setminus \{e_t\}$ of the first quiver would have the same label in the sum quiver: $\vec{\varphi}_1(e) = \vec{\varphi}_\Sigma(e)$. Analogously for the second quiver, $\forall e \in E_2 \setminus \{e_t\}, \vec{\varphi}_2(e) = \vec{\varphi}_\Sigma(e)$. And the label of the trivial edge e_t would be the sum of labels of the initial quivers: $\vec{\varphi}_\Sigma(e_t) = \vec{\varphi}_1(e_t) + \vec{\varphi}_2(e_t)$.

Predictions of the quivers are computed by Eq. 16. Now, putting everything together, we can write the sum of predictions as follows:

$$\vec{Q}_1(X) + \vec{Q}_2(X) = \sum_{e \in E_{1,X}} \vec{\varphi}_1(e) + \sum_{e \in E_{2,X}} \vec{\varphi}_2(e)$$

$$= \left(\sum_{e \in E_{1,X} \setminus \{e_t\}} \vec{\varphi}_1(e) + \sum_{e \in E_{2,X} \setminus \{e_t\}} \vec{\varphi}_2(e) \right) + \left(\vec{\varphi}_1(e_t) + \vec{\varphi}_2(e_t) \right)$$

$$= \sum_{e \in E_\Sigma \setminus \{e_t\}} \vec{\varphi}_\Sigma(e) + \vec{\varphi}_\Sigma(e_t) = \vec{Q}_\Sigma(X), \forall X \subseteq M$$

The requirement of two quiver $(L_1, E_1, \vec{\varphi}_1), (L_2, E_2, \vec{\varphi}_2)$ having only one common intent $L_1 \cap L_2 = \{\emptyset''\}$ might seem too restrictive. However, this often happens in practice for big data machine learning problems as the big datasets contain massive amount of possible intents $\mathbb{L}$. Much greater than the number of intents in both quiver: $|L_1 \cup L_2| \ll |\mathbb{L}|$.

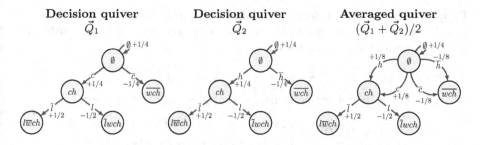

Fig. 4. Example of averaging arrow-based decision quivers.

4.4 Conversion of Target-Based Quivers to Arrow-Based Quivers

The previous subsections introduced target-based and arrow-based decision quivers. They showed that one can construct target-based decision quivers and one can sum arrow-based decision quivers. This section introduces "differentiation" operator $\delta : \dot{Q} \mapsto \vec{Q}$ that allows converting target-based decision quivers to arrow-based ones.

The desirable property of differentiation δ operator is that it should not affect the training objects predictions of the quivers:

$$\text{Given } \dot{Q}, \qquad \delta : \dot{Q} \mapsto \vec{Q}, \text{ s.t. } \dot{Q}(g') = \delta(\dot{Q})(g') = \vec{Q}(g'), \qquad \forall g \in G \qquad (19)$$

It should be noted that, given an arbitrary target-based decision quiver $\dot{Q} = (L, E, \dot{\varphi})$, we cannot define a differentiation δ operator that would not affect the quiver predictions for any given description $X \subseteq M$. This is due to the fact that the training data can contain implications that would not hold true for the test data. And difference generators of the quiver can be reflect such implications.

Example 1. Consider the example of two decision quivers from Fig. 3. Specifically, their intents $\emptyset$ and ch, and difference generators $(\emptyset, ch, c)$ and $\emptyset, ch, h)$. Both quivers start with prediction $\dot{\varphi}(\emptyset) = \vec{\varphi}((\emptyset, \emptyset, \emptyset) = 1/4$. Then, the target-based quiver predicts $1/2$ for intent ch, that is $1/4$ higher than the start prediction: $\dot{\varphi}(ch) = 1/2 = \dot{\varphi}(\emptyset) + 1/4$. The arrow-based decision quiver reflects this difference with the labels for two difference generators: $\vec{\varphi}((\emptyset, ch, c)) + \vec{\varphi}((\emptyset, ch, h)) = 1/8 + 1/8 = 1/4$. Now, the description of test object *sea snake* contains attribute c but no attribute h. Therefore, while making a prediction, the target-based quiver will follow the edge $(\emptyset, ch, c)$ and change its prediction by $1/4$: from $\dot{\varphi}(\emptyset) = 1/4$ to $\dot{\varphi}(ch) = 1/2$. And the arrow-based quiver will only change its prediction by $\vec{\varphi}((\emptyset, ch, c)) = 1/8$. Thus, two quivers would result in different predictions.

It is impossible to define what implication will be falsified on the test data. However, decision trees avoid these problem because of their dichotomic and tree properties.

Proposition 4. *Given a target-based decision tree $\dot{Q} = (L, E, \dot{\varphi})$, its differentiated version $\delta(\dot{Q})$ gives the same predictions for any input $X \subseteq M$ if differentiation operator δ defined as follows:*

For $\dot{Q} = (L, E, \dot{\varphi})$, $\dot{Q}(X) = \delta(\dot{Q})(X), \forall X \subseteq M$ if:

(L, E) *is a description tree, and* $\delta(\dot{Q}) = (L, E, \vec{\varphi})$, *where*

$$\vec{\varphi}((B_1, B_2, D)) = \begin{cases} \dot{\varphi}(\emptyset), & \text{if } B_1 = B_2 = D = \emptyset, \\ \dot{\varphi}(B_2) - \dot{\varphi}(B_1), & \text{otherwise} \end{cases} \tag{20}$$

Proof. The proof, although in different notation, is given in [8].

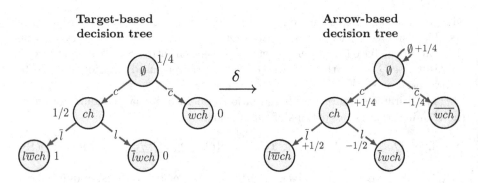

Fig. 5. Differentiation example. The arrow-based decision quiver on the right can be obtained by "differentiating" the target-based decision tree on the left.

With the introduced differentiation and summation operators we can merge ensembles of decision trees into a single decision quiver by 1) differentiating all decision trees of an ensemble, and 2) summing up these differentiated decision trees.

5 Experiments

5.1 Datasets

For this study we selected 14 real-world binary datasets from LUCS-KDD repository[1].

For all provided datasets we computed concept lattices and then selected subsets of the most stable concepts of various sizes. We use only a half of the

---

[1] Coenen, F. (2003), The LUCS-KDD Discretised/normalised ARM and CARM Data Library, http://www.csc.liv.ac.uk/~frans/KDD/Software/LUCS_KDD_DN/, Department of Computer Science, The University of Liverpool, UK.

Table 2. The description of contexts used in the experiments

id	context	# rows $\mid G \mid$	# columns $\mid M \mid$	# connections $\mid I \mid$	density $\frac{\mid I \mid}{\mid G \times M \mid}$
1	zoo	101	43	1717	0.40
2	iris	150	20	750	0.25
3	wine	178	69	2492	0.20
4	glass	214	49	2140	0.20
5	heart	303	53	4236	0.26
6	ecoli	336	35	2688	0.23
7	dematology	366	50	4750	0.26
8	breast	699	21	6974	0.48
9	pima	768	39	6912	0.23
10	anneal	898	74	12847	0.19
11	ticTacToe	958	30	9580	0.33
12	flare	1389	40	15279	0.28
13	led7	3200	25	25600	0.32
14	pageBlocks	5473	47	60203	0.23

datasets from the repository as the omitted ones result in too many concepts, thus making computation of the whole lattice infeasible. Table 2 provides information about the contexts used in the experiments.

5.2   Sizes of Difference Generators

In practice we often want to visualize the lattice of intents while presenting only the necessary attributes. In this subsection we measure to what extent difference keys allow us to shorten the labels of diagrams.

For each dataset from Table 2 we compute the whole concept lattice and estimate the stability lower bound for each concept. Then we select 10, 30, 100, and most stable concepts and compute the intent labels for the line diagram. We consider two types of labels: 1) labels drawn for each intent (i.e. attributes of an intent, that belong to no smaller intent), and 2) labels drawn between two connected intents (i.e. attributes of an intent, that do not belong to each of the preceding intents). Finally, for each label we calculate the cardinality of key of this label divided by the cardinality of the label itself.

The results are shown in Fig. 6. We see that the ratio between the lengths of the difference key and the difference often lies between 80 and 100%. That is, the keys of difference are not always shorter than the original differences. However, for some datasets and small number of selected most stable concepts, the ratio reaches 60%. Therefore, replacing intent differences by the keys of these differences can reduce the total size of attribute subsets given as labels of diagram edges up to 40%.

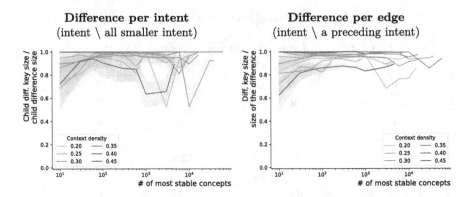

Fig. 6. Ratios between the lengths of the keys of differences and the lengths of the differences themselves. To the left: each difference consists of intent attributes that do not belong to smaller intents. To the right: each difference consists of intent attributes that do not belong to one of its preceding intents.

5.3   Summation of Decision Trees

Decision Quivers allow us to sum many decision trees in one model with no changes in the overall predictions. Here we test the summation of quivers empirically.

We test experimentally the correctness of summation operation by comparing the predictions of a random forest model and the arrow-based decision quiver constructed from this random forest. Constructing the arrow-based decision quiver from a random forest consists of three steps: 1) differentiation of all decision trees of the random forest, 2) summation of differentiated decision trees into one arrow-based decision quiver; and 3) division of the resulting decision quiver by the number of trees. That is, when random forest averages predictions of its decision trees, we "average" the decision trees themselves.

To make experiments closer to practice, we run them on the numerical (non-binarized) versions of datasets of Table 2 from UCI repository. We only take classification dataset from the table, therefore the set of outcomes Y for each dataset represents the list of probabilities of an object belonging to a specific class. For the sake of efficiency, we only consider the contexts with less than 1000 objects. For each classification dataset from the table, we make 50 random splits of the dataset to disjoint "train" and "test" subsets, where the test subset contains 20% of rows. Then, for each train-test split we fit a random forest of 100 decision trees on the train subset of the dataset. We proceed by constructing a set of arrow-based decision quivers from the first $k = 1, 2, 3, \ldots, 100$ decision trees of the forest. Finally, we use the test subset of the dataset to compute the mean average difference between the averaged predictions of the first k decision trees and the predictions of the arrow-based decision quiver constructed by averaging these k decision trees.

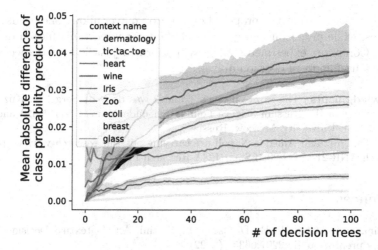

Fig. 7. The mean absolute difference between predictions of a random forest and an arrow-based decision quiver obtained by summing decision trees of the forest.

Figure 7 presents the obtained results. It can be seen that the difference between the probabilities is not null and grows with the number of decision trees. However, the difference remains small and does not exceed 5%. On average for our 12 datasets, the class probability predicted by decision quiver differs from the class probability predicted by the corresponding random forest by no more than 5%.

The difference in predictions appears because of two factors. First, as it is shown in Proposition 3, the sum of two quivers can give different predictions if these quiver have more than one common intent. Rarely, different decision trees of a random forest find the same intents while fitting. Thus, such decision trees will not satisfy the conditions of Proposition 3.

The second factor is the use of floating-point numbers. We run the experiments on numerical datasets, therefore each binary attribute used by a decision tree is constructed by comparing a real value x_i of i-th element in description x with real-valued threshold θ. We cannot perfectly fit to the precision of floating-point threshold used by sci-kit learn package. Thus, rarely, when the difference between x_i and θ is extremely small (e.g. 10^{-8}), our implementation can consider $x_i \le \theta$ as True, and sci-kit learn implementation would say the opposite. Overall, this discrepancy is of engineering nature and can be considered as a measurement error.

6   Conclusion

We have introduced and studied description quivers as compact representation of concept lattices and respective ensembles of decision trees. We have studied some properties of description quivers and tried to justify their use for describing state-of-the-art symbolic machine learning models based on decision trees.

We have shown that the proposed description quiver allows one to fuse decision trees, while proposing a way to select the most important rules in decision models. Computer experiments that we performed justified our expectations about the usefulness of decision quivers.

Acknowledgments. The work on Sects. 1 and 2 was done by Sergei O. Kuznetsov under support of the Russian Science Foundation under grant 22-11-00323 and performed at HSE University, Moscow, Russia.

Egor Dudyrev and Amedeo Napoli are carrying out this research work as part of the French ANR-21-CE23-0023 SmartFCA Research Project.

References

1. Aldinucci, T., Civitelli, E., Di Gangi, L., Sestini, A.: Contextual Decision Trees. arXiv preprint arXiv:2207.06355 (2022)
2. Bělohlávek, R., De Baets, B., Outrata, J., Vychodil, V.: Characterizing trees in concept lattices. Internat. J. Uncertain. Fuzziness Knowl. Based Syst. **16**, 1–15 (2008)
3. Bělohlávek, R., De Baets, B., Outrata, J., Vychodil, V.: Inducing decision trees via concept lattices. Int. J. Gen Syst **38**(4), 455–467 (2009)
4. Breiman, L.: Random forests. Mach. Learn. **45**, 5–32 (2001)
5. Breiman, L., Friedman, J.H., Olshen, R.A., Stone, C.J.: Classification and Regression Trees. Wadsworth, New York (1984)
6. Buzmakov, A., Dudyrev, E., Kuznetsov, S.O., Makhalova, T., Napoli, A.: Experimental study of concise representations of concepts and dependencies. In: Cordero, P., Krídlo, O. (eds.) Proceedings of the Sixteenth International Conference on Concept Lattices and Their Applications (CLA 2022), pp. 117–132. CEUR Workshop Proceedings 3308, CEUR-WS.org (2022)
7. Dudyrev, E., Kuznetsov, S.O.: Decision concept lattice vs. decision trees and random forests. In: Braud, A., Buzmakov, A., Hanika, T., Le Ber, F. (eds.) ICFCA 2021. LNCS (LNAI), vol. 12733, pp. 252–260. Springer, Cham (2021). https://doi.org/10.1007/978-3-030-77867-5_16
8. Dudyrev, E., Kuznetsov, S.O.: Summation of Decision Trees. In: Kuznetsov, S.O., Napoli, A., Rudolph, S. (eds.) Proceedings of the 9th International Workshop FCA4AI co-located with IJCAI 2021, pp. 99–104. CEUR Workshop Proceedings 2972, CEUR-WS.org (2021)
9. Ganter, B., Wille, R.: Formal Concept Analysis. Springer, Berlin (1999). https://doi.org/10.1007/978-3-030-77867-5
10. Hanika, T., Hirth, J.: Conceptual Views on Tree Ensemble Classifiers. arXiv preprint arXiv:2302.05270 (2023)
11. Kuznetsov, S.O.: Machine learning and formal concept analysis. In: Eklund, P. (ed.) ICFCA 2004. LNCS (LNAI), vol. 2961, pp. 287–312. Springer, Heidelberg (2004). https://doi.org/10.1007/978-3-540-24651-0_25
12. Strecht, P.: A survey of merging decision trees data mining approaches. In: Proceedings of the 10th Doctoral Symposium in Informatics Engineering (DSIE 2015), pp. 36–47 (2015)
13. Wille, R.: Line diagrams of hierarchical concept systems. Knowl. Organ. **11**(2), 77–86 (1984). Nomos Verlagsgesellschaft mbH & Co. KG

Examples of Clique Closure Systems

Elias Dahlhaus[ID] and Bernhard Ganter[(✉)][ID]

Ernst-Schröder–Zentrum für Begriffliche Wissensverarbeitung, Darmstadt, Germany
esz@mathematik.tu-darmstadt.de

Abstract. We give a complete list of lattices associated to closure systems generated by four maximal cliques and generalize a known characterization to uniform hypergraphs.

Keywords: graph · hypergraph · clique · closure system · coatomistic lattice · Formal Concept Analysis

1 Introduction

The following notes arose from a discussion that the authors had during a Dagstuhl meeting on *"Concept Lattice Based Topological Data Analysis and Reasoning"*. The first author presented a result characterizing configurations of maximal cliques in graphs. Later we learned that other authors had already dealt with this topic in detail, but we did not find any example list obtained from such a description. Therefore, we describe here all configurations of up to four maximum cliques and explain how our list was created. The same methods could be used to expand the list to five and six maximal cliques.

From the point of view of Formal Concept Analysis, it is interesting to generalize the result to hypergraphs. Hypergraphs generalize graphs in that edges may consist of more than two nodes. In general, it is not required that all edges have the same cardinality. Hypergraphs for which this is the case are called **uniform**. A hypergraph is called k-**uniform** if each of its edges consists of exactly k nodes. Thus 2-uniform hypergraphs are just the ordinary graphs. A **clique** in a k-uniform hypergraph is a set of nodes, each k-elementary subset of which is a hyperedge.

From each formal context, one immediately obtains two hypergraphs, that of attribute extents and that of object intents. In the following investigation, however, a formal context is derived in a different way: The objects are the nodes of the (hyper-) graph, and the attributes are the maximal cliques.

2 Maximal Cliques

A **clique** in a graph (V, E) is a subset $C \subseteq V$ with the property that every two distinct vertices in C are joined by an edge of E. A **maximal clique** is a clique that is not contained in any other clique. From the Axiom of Choice one concludes that every clique is contained in some maximal clique.

D. Dürrschnabel and D. López Rodríguez (Eds.): ICFCA 2023, LNAI 13934, pp. 143–151, 2023.
https://doi.org/10.1007/978-3-031-35949-1_10

For any graph with vertex set V the system $\mathcal{H}$ of all maximal cliques has three obvious properties:

MC1 $\mathcal{H}$ **covers** V, i.e., every vertex in V is contained in some $H \in \mathcal{H}$,
MC2 no $H_1 \in \mathcal{H}$ is properly contained in some $H_2 \in \mathcal{H}$, and
MC3 whenever a subset $R \subseteq V$ with $|R| \geq 2$ is not contained in any $H \in \mathcal{H}$,
 then R contains a two-element set with the same property.

Indeed, it is easy to show that the conditions MC1–MC3 characterize systems of maximal cliques of graphs, and that the unique graph on V with $\mathcal{H}$ as its family of maximal cliques has the edge set

$$\{\{u, v\} \mid u, v \in V, u \neq v, \{u, v\} \subseteq H \text{ for some } H \in \mathcal{H}\}.$$

Here we use another characterization of such families of maximal cliques, which can be found in [1], where it is attributed to [3]:

Proposition 1 (Gilmore's condition). *A set $\mathcal{H}$ of subsets of a finite set V is the set of all maximal cliques of some graph (V, E) if and only if $\mathcal{H}$ satisfies MC1, MC2, and*

MC4 *for any three sets A, B, C in $\mathcal{H}$ there is some $D \in \mathcal{H}$ such such that*

$$(A \cap B) \cup (A \cap C) \cup (B \cap C) \subseteq D.$$

Proposition 1 can be generalized from graphs to uniform hypergraphs. The conditions above then need modification, since the definition of a clique must be adapted: A clique in a k-uniform hypergraph (V, E) is a set $C \subseteq V$ with the property that every k distinct elements of C are joined by an edge. Thereby every vertex set with less than k elements automatically is a (trivial) clique. Instead of MC1, then

MC1. k Every $k - 1$-element set of vertices is contained in some $H \in \mathcal{H}$

is needed.
    Note that for $k \geq 2$ any given set $\mathcal{H}$ of sets **induces** a k-uniform hypergraph $(V := \bigcup \mathcal{H}, E)$, the hyperedges of which are exactly the k-element subsets of V which are contained in some $H \in \mathcal{H}$, formally

$$e \in E : \iff |e| = k \text{ and } e \subseteq H \text{ for some } H \in \mathcal{H}.$$

Theorem 1. *Suppose $k \geq 2$. A set $\mathcal{H}$ of subsets of a finite set V is the set of all maximal cliques of some k-uniform hypergraph on V if and only if $\mathcal{H}$ satisfies MC1.k, MC2, and*

MC4.k *For each choice of $A_0, A_1 \ldots, A_k \in \mathcal{H}$ there is some $D \in \mathcal{H}$ such that*

$$\bigcup_{i=0}^{k} \bigcap_{j \neq i} A_j \subseteq D.$$

Proof. The proof is done by induction, with Proposition 1 as anchoring (for $k = 2$).

To prove $\Leftarrow$, we show that every clique is contained in some $H \in \mathcal{H}$. So let C be a clique of the k-uniform hypergraph (V, E) induced by $\mathcal{H}$, and choose $H \in \mathcal{H}$ so that H maximally intersects C. If $C \subseteq H$, then we are done. If otherwise, then we may assume that $|C \cap H| \geq k$, since the smaller cases are trivial. Pick some vertex $v \in C \setminus H$, let $\mathcal{N}$ be the set of those sets in $\mathcal{H}$ which maximally intersect H while containing v, and define

$$\mathcal{N}'' := \{N \cap H \mid N \in \mathcal{N}\}.$$

We claim that $\mathcal{N}''$ satisfies condition MC4.$(k-1)$. To show this we must prove that for each choice of k elements $N_0, \ldots, N_{k-1} \in \mathcal{N}$ the set

$$\bigcup_{i=0}^{k-1} \bigcap_{j \neq i} (N_j \cap H) \tag{1}$$

is contained in some element of $\mathcal{N}''$. But note that

$$\bigcup_{i=0}^{k-1} \bigcap_{j \neq i} (N_j \cap H) = H \cap \bigcup_{i=0}^{k-1} \bigcap_{j \neq i} N_j.$$

Since $\mathcal{H}$ fulfills condition MC4.k, we know that there is some $D \in \mathcal{H}$ with

$$\bigcup_{i=0}^{k} \bigcap_{j \neq i} N_j \subseteq D \quad \text{for } N_k := H.$$

There may be several such D. We choose one that maximally intersects H. D contains $\bigcap_{j \neq k} N_j$ and therefore v, thus $D \in \mathcal{N}$ and $D \cap H \in \mathcal{N}''$. Also D contains for each $i \leq k-1$ the set

$$\bigcap_{i \neq j \leq k} N_j = \bigcap_{i \neq j \leq k-1} (N_j \cap H),$$

and thus contains (1), which proves the claim. Form the induction hypothesis we conclude that $\mathcal{N}''$ is the set of maximal ciques of some $(k-1)$-uniform hypergraph.

Now consider a set e of $k-1$ points of $C \cap H$. Together with v they form a k-hyperedge $e \cup \{v\}$ (since C is a clique in the k-hypergraph induced by $\mathcal{H}$), and thus there must be some $M \in \mathcal{H}$ containing this edge. There may be several such sets, so pick one that maximally intersects H. Since $v \in H$ we have that $M \in \mathcal{N}$, thus $M \cap H \in \mathcal{N}''$, which shows that e is an hyperedge of the $(k-1)$-uniform hypergraph. Since the choice of e was arbitrary, we conclude that all $(k-1)$-subsets of $C \cap H$ belong to that hypergraph, and that $\{v\} \cup (C \cap H)$ is a clique. This clique must be contained in some maximal clique, i.e., in some element $N \cap H$ of $\mathcal{N}''$. But then N is an element of $\mathcal{H}$ that contains all of $C \cap H$ and, in addition, the vertex v. This contradicts the assumption that H maximally intersects C, and therefore shows that C must be contained in some element of $\mathcal{H}$.

Proving $\Rightarrow$ is easier. It is obvious that MC1.k and MC2 must hold for the system $\mathcal{H}$ of maximal cliques of a k-uniform hypergraph. It remains to show MC4.k. Let $A_0, \ldots, A_k \in \mathcal{H}$ and let $v_1, \ldots, v_k$ be vertices, each of which is in at least k of the A_i. Then each v_j is missing at most one of the A_i, and there must be one of the A_i which contains all of $v_1, \ldots, v_k$. This shows that each k-element subset of $\bigcup_{i=0}^{k} \bigcap_{j \neq i} A_j$ is a hyperedge, so that $\bigcup_{i=0}^{k} \bigcap_{j \neq i} A_j$ is a clique and must be contained in some maximal clique $D \in \mathcal{H}$. $\square$

3   Closed Cliques

The set of all cliques of a graph is usually not closed under intersection. Those cliques which are intersections of maximal cliques will be called **closed cliques**. The **clique closure system** of a graph consists of all closed cliques plus the base set V, although that normally is not a clique. Ordered by $\subseteq$ this closure system is a complete lattice. Let us recall from lattice theory that a lattice is **coatomistic** if every element is the meet of coatoms, i.e., of lower neighbors of the largest element.

Brucker and Gély [1] have studied these lattices, in particular under the aspect[1] of being *crown-free*. They did not explicitly state the following characterization, which can however be derived from Proposition 1.

Proposition 2. *A closure system covering a finite set V is a clique closure system if and only if, as a lattice, it is coatomistic and does not contain an 8-element Boolean sublattice with top element V.*

Proposition 2 can be generalized to k-uniform hypergraphs.

Theorem 2. *Let $k \geq 2$. A closure system satisfying MC1.k on a finite set V is the clique closure system of a k-uniform hypergraph iff it is coatomistic and does not contain a Boolean lattice with $k + 1$ atoms and with largest element V as a $\vee$-subsemilattice.*

Proof. By definition is every clique in a clique closure system an intersection of maximal cliques, so these must be the coatoms and MC.2 holds. MC1. k was assumed.

It remains to check MC4. k. This condition is violated if and only if there are maximal cliques $A_0, \ldots, A_k$ such that $\bigcup_i \bigcap_{j \neq i} A_j$ is not contained in any maximal clique. Let $C_i := \bigcap_{j \neq i} A_j$ for all i. Whenever $T \subseteq \{0, \ldots, k\}$, then

$$\bigvee_{i \in T} C_i \subseteq A_j \iff j \notin T.$$

---

[1] See also Järvinen and Radeleczki [4] for other aspects.

All these joins therefore are different and form a ∨-subsemilattice isomorphic to a Boolean lattice. Condition MC4.k is violated only if there is no coatom containing all C_i, i.e., if the join of all C_i is V.

Conversely if a coatomistic lattice with top element ⊤ contains a Boolean ∨-subsemilattice with the same top element, then let B_i, C_i be the atoms and coatoms of this subsemilattice, and let the numbering be such that $B_i \subseteq C_j \iff i \neq j$. Each C_i must be contained in some lattice coatom A_i. Any intersection $\bigcap_{j \neq i} A_j$ of k of these coatoms contains an atom, B_i. A coatom containing all the intersections therefore must contain all sets B_i, but their least common upper bound is ⊤. Therefore $A_0, \ldots A_k$ would violate MC4.k.   □

Remark. The clique closure system of the 3-uniform hypergraph

$$(\{0, \ldots, 5\}, \{\{0, 1, 2\}, \{2, 3, 4\}, \{4, 5, 0\}\})$$

contains an 8-element Boolean sublattice with top element $\{0, \ldots, 5\}$, and thus cannot be represented by a graph clique closure system according to Proposition 2. A much simpler example is given by the 3-uniform hypergraph $(\{0, 1, 2\}, \emptyset)$ with no hyperedges at all. Its (trivial) maximal cliques are the three 2-element subsets. The clique closure system is isomorphic to the 8-element Boolean lattice.

4   Examples

Closure systems with isomorphic lattices can be different, but, as also mentioned in [1], they can be reduced to a canonical minimal one, at least in the finite case, by removing superfluous vertices. For example, if two adjacent vertices are contained in exactly the same cliques, then they can be merged to one without changing the hierarchy of the closed cliques. Doing so repeatedly reduces the number of vertices, so that eventually each closed clique contains at most one vertex which is not contained in any smaller clique. But even such a vertex v may be removed, except when there is a maximal clique containing all other vertices in that clique. It turns out that exactly one vertex remains for each closed clique that is join-irreducible in the closure system.

To get an overview of small examples, we proceeded as follows: If a graph has exactly h maximal cliques, then by Proposition 2 these cliques generate a coatomistic closure system with exactly h coatoms. Closure systems with a small number of infimum-irreducible elements are easy to generate using algorithms from Formal Concept Analysis, see [2]. The desired examples can then be filtered out. For each such example there is a compact representation, given by the so-called standard context of the associated lattice.

For $h \leq 2$ there are only trivial examples. There are exactly four coatomistic lattices with 3 coatoms. One of them is Boolean and therefore not the lattice of a clique closure system. The other three are, and we show for each its standard context, a lattice diagram, and a minimal graph representing that clique closure system. A more detailed reading instruction is given after Example 4.20 below.

3.1			

3.2			

3.3			

In the first two examples, the graph is disconnected. It is not difficult to describe how the lattice can then be composed of the components.

Up to isomorphism, there are 126 lattices with exactly four meet-irreducible elements. 50 of them are coatomistic, and 20 of these satisfy the conditions from Proposition 2. Seven of these 20 examples (numbered 4.1–4.7) have disconnected graphs, see Fig. 1.

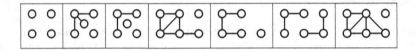

Fig. 1. Seven disconnected graphs for clique closure systems

That leaves 13 examples, listed below, for a complete census of minimal clique closure systems with four maximal cliques.

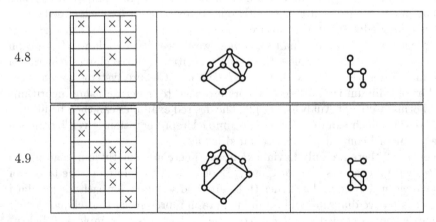

4.8			

4.9			

4.10

×		×	×
			×
×	×	×	
		×	
	×		
×			

4.11

×	×		×
			×
×		×	
		×	
×	×		
	×		

4.12

	×		×
			×
×		×	
		×	
×	×		

4.13

		×	×
	×		×
×		×	
×	×		

4.14

×			×
			×
×		×	
		×	
×	×		
	×		

4.15

×	×		×
			×
×	×	×	
×		×	
		×	
	×		

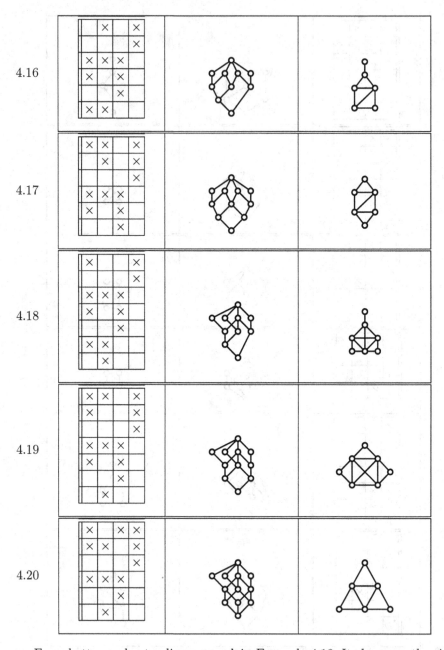

For a better understanding we explain Example 4.18. It shows on the right a graph with seven nodes and eleven edges. The seven nodes correspond to the seven objects in the formal context on the left. The attribute extents of the context are the four maximal cliques of the graph, one of which consists of four nodes, two of three nodes, and one of two nodes. Consequently, the formal context has one column with four incidence crosses, two columns with three each, and one with only two crosses. The line diagram in the middle shows the concept lattice of this formal context. It has seven join irreducible concepts, corresponding to the nodes of the graph. The four meet irreducible concepts all are coatoms, because they represent the maximal cliques, which are necessarily incomparable.

References

1. Brucker, F., Gély, A.: Crown-free lattices and their related graphs. Order **28**, 443–454 (2011)
2. Ganter, B., Obiedkov, S.: Conceptual Exploration, Springer, Berlin (2016). https://doi.org/10.1007/978-3-662-49291-8
3. Gilmore, P.C.: Families of sets with faithful graph representation. J. Watson Research Center, Yorktown Heights, Technical report (1962)
4. Järvinen, J., Radeleczki, S.: Tolerances induced by irredundant coverings. Fund. Inform. **137**(3), 341–353 (2015)

On the Maximal Independence Polynomial of the Covering Graph of the Hypercube up to $n{=}6$

Dmitry I. Ignatov[✉] [ID]

HSE University, Moscow, Russia
digntov@hse.ru

Abstract. There are well-known problems in extremal set theory that can be formulated as enumeration of the maximal independent sets or counting their total number in certain graphs. Here we provide an FCA-based solution on the number of maximal independent sets of the covering graph of a hypercube. In addition, we consider the related maximal independence polynomials for n up to 6, and prove several properties of the polynomials' coefficients and the corresponding concept lattices.

Keywords: Close-by-one · maximal independent set · hypercube · maximal independence polynomial

1 Introduction

Independent sets are important for many applications of Graph Theory in Computer Science, for example, in Coding Theory [2] or Network Analysis [1]. An independent set (IS) in a graph is its subset of vertices such that any two of them are not connected. There are several algorithms for enumeration of all independent sets or for searching all maximal or maximum independent sets [15,19]. The last two problems are considered in extremal combinatorics. Maximal independent sets (MIS)[1] are independent sets that cannot be extended by adding new vertices, while maximum independent sets are the largest possible independent sets for a given graph.

Knowledge of the concrete formulas for the number of independent sets of various graphs is limited and sometimes we know only the first few values[2] or their asymptotics only w.r.t. the size of an input graph. In this paper, as our input graph, we consider a hypercube. Its asymptotic behaviour is known due to a series of results [3,4,10,11]. More structural information is contained in the so-called independence polynomials where their coefficients equal to the number of independent sets of the size equals to the related variable's degree [17].

In this paper, we obtain structural information on the size of maximal independent sets and consider the so-called maximal independence polynomials [8] of a hypercube.

---

[1] We use MIS both for the plural and singular forms depending on context.

[2] https://oeis.org/A284707.

© The Author(s), under exclusive license to Springer Nature Switzerland AG 2023
D. Dürrschnabel and D. López Rodríguez (Eds.): ICFCA 2023, LNAI 13934, pp. 152–165, 2023.
https://doi.org/10.1007/978-3-031-35949-1_11

In spite of standard algorithms for the problem of enumeration and counting of maximal independent sets, we use the problem formulation in terms of concept lattices, where any maximal independent set of a graph corresponds to a formal concept of the context obtained from the complement of the adjacency matrix of this graph.

The paper consists of the following sections. Section 2 states the studied problem and overviews related facts. Section 3 briefly discusses a modification of CLOSE BY ONE algorithm for maximal independent set counting and enumeration. Section 4 summarises all the obtained results. Finally, Sect. 5 concludes the paper.

2   Problem Statement and Basic Facts

2.1   Graphs and Independent Sets

A *cube graph* is defined as follows: the vertices of Q_n are the 0, 1-vectors of length n, with two adjacent vertices if they differ in a single coordinate. Equivalently, we can consider each vector $(s_1, s_2, \ldots, s_n)$ as the subset S of $[n] = \{1, 2, \ldots, n\}$ where $S = \{i \in [n] \mid a_i = 1\}$.

If we consider the power set $2^{[n]}$ equipped with set inclusion relation $\subseteq$, then Q_n is the covering graph of the power set lattice.

The *complement graph* $\Gamma^c = (V, E^c)$ of a graph $\Gamma = (V, E)$ is given by $\{u, v\} \in E^c \iff \{u, v\} \notin E$. It is known that (maximal) cliques (or (maximal) complete subgraphs) of a graph are (maximal) independent sets of its complement graph.

In what follows we will denote the adjacency matrix of the complement of the hypercube graph as $A(Q_n)$.

The *independence number* of a graph Γ [16] is the cardinality of the largest independent vertex set and is denoted by $\alpha(\Gamma)$.

It is known that the maximum size independent set in the hypercube of dimension n is $\alpha(Q_n) = 2^{n-1}$ [16] and this is also the largest exponent in the *independence polynomial* of Q_n.

For example, the independence polynomial of Q_4 has the following form:

$$2x^8 + 16x^7 + 56x^6 + 128x^5 + 228x^4 + 208x^3 + 88x^2 + 16x + 1 \ .$$

In general, the exponent of each monomial in an independence polynomial corresponds to the size of an independent set, while the monomial coefficient equals the number of IS of that size.

The general form of the *maximal independence polynomial* of a graph $\Gamma = (V, E)$ is given by

$$I_{max}(\Gamma, x) = \sum_{k=0}^{\alpha(\Gamma)} b_k x^k, \text{ where}$$

k is the size of a maximal independent set and b_k is the number of MIS of that size. Our goal here is to obtain these polynomials for n up to 6 and study their properties and related structures.

The asymptotic behaviour of the number of MIS in Q_n is $mis(Q_n) \sim 2n \cdot 2^{2^n/4}$ [11].

2.2 Formal Concept Analysis and Symmetric Concepts

To further proceed with the problem statement, we need to introduce several definitions from Formal Concept Analysis [7], an applied branch of modern Lattice Theory. We reproduce basic definitions from our tutorial [9], for more details see also textbook [6].

A *formal context* $\mathbb{K} = (G, M, I)$ consists of two sets G and M and a relation I between G and M. The elements of G are called the *objects* and the elements of M are called the *attributes* of the context. The notation gIm or $(g, m) \in I$ means that the object g has attribute m.

Two special types of context defined on any set S are used in the remaining sections: the *nominal scale* $\mathbb{N}_S = (S, S, =)$ and the *contranominal scale* $\mathbb{N}_S^c = (S, S, \neq)$.

For $A \subseteq G$ and $B \subseteq M$, let

$$A' := \{m \in M | (g, m) \in I \text{ for all } g \in A\}$$

$$B' := \{g \in G | (g, m) \in I \text{ for all } m \in B\}.$$

These operators are called *derivation operators* or *concept-forming operators* for $\mathbb{K} = (G, M, I)$.

Proposition 1. *Let* (G, M, I) *be a formal context, for subsets* $A, A_1, A_2 \subseteq G$ *and* $B \subseteq M$ *we have*

1. $A_1 \subseteq A_2 \Rightarrow A_2' \subseteq A_1'$ *(antimonotony of* $'$*)*,
2. $A_1 \subseteq A_2 \Rightarrow A_1'' \subseteq A_2''$ *(monotony of* $''$*)*,
3. $A \subseteq A''$ *(extensity of* $''$*)*,
4. $A' = A'''$ *(hence,* $A'''' = A''$*, i.e. idempotency of* $''$*)*,
5. $(A_1 \cup A_2)' = A_1' \cap A_2'$,

Similar properties hold for subsets of attributes.

For $\mathbb{K} = (G, M, I)$, the operators $(\cdot)'': 2^G \to 2^G$, $(\cdot)'': 2^M \to 2^M$ are closure operators, i.e. idempotent, extensive, and monotone.

A *formal concept* of a formal context $\mathbb{K} = (G, M, I)$ is a pair (A, B) with $A \subseteq G$, $B \subseteq M$, $A' = B$ and $B' = A$. The sets A and B are called the extent and the intent of the formal concept (A, B), respectively. The *subconcept-superconcept relation* is given by $(A_1, B_1) \leq (A_2, B_2)$ iff $A_1 \subseteq A_2$ ($B_2 \subseteq B_1$).

The set of all formal concepts of a context $\mathbb{K}$ together with the order relation $\leq$ forms a complete lattice called the *concept lattice* of $\mathbb{K}$ and denoted by $\underline{\mathfrak{B}}(\mathbb{K})$.

In what follows, we consider the following graph-induced context $\mathbb{K}(\Gamma) = (V, V, I)$ of a graph $\Gamma = (V, E)$ where $vIu \iff \{v, u\} \in E$. Since I is a symmetric relation ($vIu \Rightarrow vIv$), we call $\mathbb{K}(\Gamma)$ as a symmetric context. We call a formal concept $\mathbb{K}(\Gamma)$ with equal extent and intent symmetric.

It is clear that symmetric formal concepts of $\mathbb{K}(\Gamma)$ are in one-to-one correspondence with maximal cliques in Γ where each extent (intent) of a symmetric concept forms the corresponding maximal clique. Moreover, the symmetric concepts of $\mathbb{K}(\Gamma)$ form an antichain (a set of pairwise incomparable elements) $\mathfrak{M}^+$ with respect to the subconcept-superconcept relation. This antichain is a subset of the set of all equally sized concepts of $\mathbb{K}(\Gamma)$, which forms a maximal antichain $\mathfrak{M}$ of $\underline{\mathfrak{B}}(\mathbb{K}(\Gamma))$.

2.3   Conceptual Representation of Q_n

Let us consider the complementary context of $\mathbb{K}(Q_n)$, i.e. $\mathbb{K}^c(Q_n) = (2^{[n]}, 2^{[n]}, (2^{[n]} \times 2^{[n]}) \setminus I)$.

In Fig. 1, the exemplary contexts $\mathbb{K}^c(Q_n)$ are given for $n = 0, 1, 2$. Note that the object names are given as integers, whose binary representation encodes the corresponding sets, e.g. $7 = 111_2$, i.e. $\{0, 1, 2\}$.

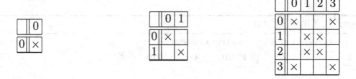

Fig. 1. Example contexts for $\mathbb{K}^c(Q_0)$, $\mathbb{K}^c(Q_1)$, and $\mathbb{K}^c(Q_2)$.

The following recursive rule can be applied for obtaining these contexts for higher n:

$$\mathbb{K}^c(Q_{n+1}) = \frac{\mathbb{K}^c(Q_n) \mid \mathbb{N}^c_{2^{[n]}}}{\mathbb{N}^c_{2^{[n]}} \mid \mathbb{K}^c(Q_n)}.$$

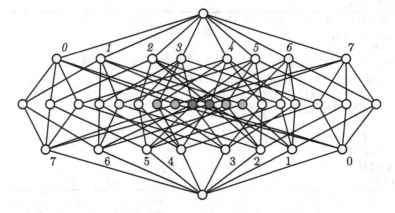

Fig. 2. The line diagram of $BV(\mathbb{K}(Q_4))$. Symmetric concepts with four-element extents (intents) are shown in grey, while those with two-element extents are in light grey.

The line diagram of $\mathfrak{B}(\mathbb{K}(Q_4))$ shown in Fig. 2. The nodes corresponding to symmetric concepts are shaded and they form the antichain $\mathfrak{M}_+ = \{(1247, 1247), (0356, 0356), (07, 07), (16, 16), (25, 25), (34, 34)\}$ and each extent (intent) of them is a MIS of Q_4.

3   Algorithm

Close-by-One (CBO) is one of the well-known algorithms for formal concept generation [12]. Due to its monotonic way of object (attribute) sets exploration and antimonotony property of $(\cdot)'$ operator, which relates extents and intents, we can modify the algorithm as follows (see Algorithm 1 and Algorithm 2): 1) for a current concept, we check whether the extent is equal to the intent and store it if any; 2) we stop each branch execution if the extent of a current concept is not a proper subset of its intent. Note that the input should be a symmetric context.

Algorithm 1. CBOSYMM

Input: $\mathbb{K} = (G, G, I)$ is a symmetric context
Output: $\mathfrak{M}^+$ is the set of all symmetric concepts of $\mathbb{K}$
 1: $\mathfrak{M}^+ := \emptyset$
 2: **for all** $g \in G$ **do**
 3:     Process($\{g\}, g, (\{g\}'', \{g\}')$)
 4: **end for**
 5: **return** $\mathfrak{M}^+$

Algorithm 2. Process($A, g, (C, D)$) with $C = A''$ and $D = A'$ and $<$ the linear order on objects

---

 1: **if** $\{h|h \in C \setminus A$ and $h < g\} = \emptyset$ **then**
 2:     **if** $C = D$ **then**
 3:         $\mathfrak{M}^+ := \mathfrak{M}^+ \cup \{(C, D)\}$
 4:     **else if** $C \subset D$ **then**
 5:         **for all** $f \in \{h|h \in G \setminus C$ and $g < h\}$ **do**
 6:             $Z := C \cup \{f\}$
 7:             $Y := D \cap \{f\}'$
 8:             $X := Y'$
 9:             Process($Z, f, (X, Y)$)
10:         **end for**
11:     **end if**
12: **end if**

The worst case time complexity of CBO is $O(|G|^2|M||L|)$, where L is the set of output concepts, and its polynomial delay is $O(|G|^3||M|)$ for a nominal scale context $(G, G, =)$. However, the complexity of CBOSYMM on a nominal scale is $O(|G|^3)$ $(\mathfrak{M}^+ = G)$ and the polynomial delay is $O(|G|^2)$ (the complexity of closure operation) since Algorithm 2 stops when $\{g\}'' = \{g\}'$. In general, CBOSYMM processes $(|L| + |\mathfrak{M}|)/2$ closed sets compared to $|L|$ where $\mathfrak{M}$ is a set of all concepts with equally sized extents and intents.

The generation protocol of CBO in a tree-like form is given in Fig. 3. Each closed set of objects (extent) can be read from the tree by following the path from the root to the corresponding node. Square bracket] means that the first prime operator has been applied after the addition of the next object g to the set A of the parent node and bracket) shows which object has been added after application of the second prime operator, i.e. between] and) one can find $(A \cup \{g\})'' \setminus (A \cup \{g\})$. A non-canonic generation can be identified by simply checking whether there is an object between] and) that is less than g w.r.t. $<$. One can note that the generation tree is traversed in a depth-first search manner.

As for CBOSYMM, whenever MIS is found, all its outgoing branches present in CBO protocol are not executed.

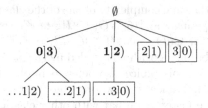

Fig. 3. The tree of CBO protocol for the context $\mathbb{K}^c(Q_2)$. Non-canonic generations are drawn in boxes. The extents (MIS) outputted by CBOSYMM are in bold.

Since Algorithm 1 processes each object separately, we are able to implement its parallel version. For counting purposes, we need to replace $\mathfrak{M}_+$ in CBOSYMM by a counter variable or by an array of counters for MIS of different sizes.

In Fig. 4 one can see the pairwise plot of the counters in CBOSYMM branches that started with a single element set (the horizontal axis) and then added a new element shown on the vertical axis. Empty cells mean that we skip the corresponding branches for pairs of connected elements of Q_n like 4 and 5.

Our implementation of CBOSYMM in Python can be found in an iPython notebook[3] on GitHub[4] along with the protocols of main experiments. We use integers to pack binary vectors representing objects (or their intents) for efficiency purposes. In addition, we use Cython[5] to execute the code on par with low-

---

[3] https://ipython.org/notebook.html.
[4] https://github.com/dimachine/CubeIndSets.
[5] https://cython.readthedocs.io/.

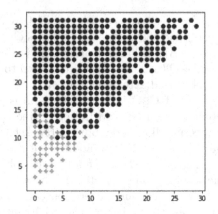

Fig. 4. The pairwise plot of the counters in CBOSYMM branches for the context $\mathbb{K}^c(Q_5)$. The + sign means non-zero values, the dot is for zero.

level language with strong typisation. Since the branches in CbO and our modification can be processed in parallel we also use Python multiprocess library[6].

We also note that the time complexity of one of the classic algorithms for MIS enumeration by Tsukiyama et al. [15] is $O(|G||I|mis(\mathbb{K}))$, which is not essentially different from that of for CBO since $|I| \le |G||M| = |G|^2$. The rationale behind the choice of an FCA-based algorithm is guided by the lattice-theoretic flavour of the considered problems; this choice also helps us in obtaining some interesting discoveries (see Sect. 4). We leave answering the question of whether an FCA-based algorithm can be faster or on par with other classic or recent algorithms for MIS enumeration for a separate study.

4   Results

4.1   Level-Union MIS

Hence, prior to computations, the number of non-zero coefficients for Q_n is not larger than 2^{n-1}, which is also equal to the lower bound on the maximal number of MIS of different sizes for a graph with 2^n vertices [5].

Our goal here is at least to indicate which coefficients are non-zeros and provide the reader with their lower bounds.

Theorem 1. *Let Q_n be a hypercube and its levelwise representation is given by the diagram*

$$\binom{[n]}{0} - \binom{[n]}{1} - \cdots - \binom{[n]}{n},$$

---

[6] https://pypi.org/project/multiprocess/.

then the number of maximal independent sets obtained as the union of its levels $\binom{[n]}{k}$[7] *is given by a recurrent relation* $a_n = a_{n-2} + a_{n-3}$ *for* $n > 2$, *where* $a_0 = 1$, $a_1 = 2$, $a_2 = 2$.

Proof. From the diagram, we can deduce that this problem is equivalent to determining the number of maximal independent sets in the chain (or the path graph) on $n + 1$ elements. Let us apply an inductive argument here. For $n = 0$ there the MIS is the single element, for $n = 1$ each path node is a MIS, for $n = 2$ two end nodes and the central node form two different MIS, respectively.

Let P_n be a path graph of length $n > 2$ on $n + 1$ nodes. Let $mis(P_n \mid k)$ denote the number of MIS of P_n that contain node k (the nodes are numbered from the left end to the right end starting from 0).

To obtain MIS in P_n containing n we can simply extend each MIS of P_{n-2} containing $n - 2$ and each MIS of P_{n-3} containing $n - 3$ by adding node n, i.e., $mis(P_n \mid n) = mis(P_{n-3} \mid n - 3) + mis(P_{n-2} \mid n - 2)$.

Similarly, $mis(P_n \mid n-1) = mis(P_{n-3} \mid n-3) + mis(P_{n-3} \mid n-4)$. However, $mis(P_{n-3} \mid n - 3) = mis(P_{n-2} \mid n - 3)$. Therefore, $a_n = mis(P_n \mid n) + mis(P_n \mid n - 1) = a_{n-2} + a_{n-3}$ since $a_{n-2} = mis(P_{n-2} \mid n - 2) + mis(P_{n-2} \mid n - 3)$ and $a_{n-3} = mis(P_{n-3} \mid n - 3) + mis(P_{n-3} \mid n - 4)$ by the disjointness of any two MIS families that contain different neighbouring nodes.

Corollary 1. *Let* Q_n *be a hypercube then the number of maximal independent sets obtained as the union of its levels* $\binom{[n]}{k}$ *has the following generation function:*

$$\frac{x^2 + 2x + 1}{1 - x^2 - x^3}.$$

Proof. Let $g(x) = \sum_{n=0}^{\infty} a_n x^n$ be the generating function of a_n. From Theorem 1 we have

$$\sum_{n=3}^{\infty} a_n x^n = \sum_{n=3}^{\infty} a_{n-2} x^n + \sum_{n=3}^{\infty} a_{n-3} x^n.$$

Further,

$$g(x) - a_0 - a_1 x - a_2 x^2 = x^2(g(x) - a_0) + x^3 g(x)$$

or

$$g(x)(1 - x^2 - x^3) = a_0 + a_1 x + (a_2 - a_0)x^2.$$

Corollary 2. *Let* Q_n *be a hypercube then the number of maximal independent sets obtained as the union of its levels* $\binom{[n]}{k}$ *is as follows:*

$$a(n) = \sum_{z \in \{\alpha, \beta, \gamma\}} \frac{1}{z^n} \frac{2 + 2z + z^2}{2z + 3z^2}, \quad where$$

---

[7] $\binom{[n]}{k}$ are subsets of $[n]$ of size k.

α *is the real root of* $x^3 + x^2 = 1$ *and* β *and* γ *are the complex conjugate roots.*

Proof. It immediately follows from Theorem 1 and the solution for the general linear third-order recurrence equation $a_n = Aa_{n-1} + Ba_{n-2} + Ca_{n-3}$ [18].

$$a_n = a_1 \sum_{z \in \{\alpha,\beta,\gamma\}} \frac{z^{-n}}{A + 2zB + 3z^2 C} - (Aa_1 - a_2) \sum_{z \in \{\alpha,\beta,\gamma\}} \frac{z^{1-n}}{A + 2zB + 3z^2 C}$$

$$- (Ba_1 + Aa_2 - a_3) \sum_{z \in \{\alpha,\beta,\gamma\}} \frac{z^{2-n}}{A + 2zB + 3z^2 C}, \text{ where}$$

α, β, γ are the roots of $Ax + Bx^2 + Cx^3 = 1$.

The value $\rho = 1/\alpha = \sqrt[3]{\frac{9+\sqrt{69}}{18}} + \sqrt[3]{\frac{9-\sqrt{69}}{18}} = 1.3247179572...$ is known as the plastic number and describes Perrin[8] and Padovan[9] sequences via similar expressions.

Proposition 2. *The asymptotic behaviour of* a_n *is*

$$a_n \sim \frac{1}{\alpha^n} \frac{2 + 2\alpha + \alpha^2}{2\alpha + 3\alpha^2} = \rho^n \frac{2\rho^2 + 2\rho + 1}{2\rho + 3}.$$

Proof. From Corollary 2 we know that α is the real root of $x^3 + x^2 = 1$, i.e. $\alpha = \frac{1}{3}(-1 + \sqrt[3]{\frac{1}{2}(25 - 3\sqrt{69})} + \sqrt[3]{\frac{1}{2}(25 + 3\sqrt{69})}) \approx 0.75488$, while β and γ are the conjugate roots $\beta \approx -0.87744 - 0.74486i$ and $\gamma \approx -0.87744 + 0.74486i$.

Since the absolute values of α and β are both greater than 1 (≈ 1.151), the reciprocals of the powers of these roots tend to 0 for large n, and we can omit the terms containing them from the asymptotic formula.

The obtained approximation is very accurate even for the beginning values in Table 1.

Table 1. a_n and its approximation

n	0	1	2	3	4	5	6	7	8	9	10
a_n	1	2	2	3	4	5	7	9	12	16	21
approximation	1.27	1.68	2.22	2.95	3.90	5.17	6.85	9.07	12.02	15.92	21.10
relative error	0.211	0.191	0.101	0.018	0.025	0.0328	0.022	0.008	0.002	0.005	0.004

Theorem 1 does not say much about the structure of the obtained solutions, however, its proof contains a direct way to enumerate and count these MIS.

---

[8] https://oeis.org/A001608.
[9] https://oeis.org/A000931.

Our Python-based implementation helped to identify the following MIS for $n = 6$ and $n = 7$ in Table 2. The sizes of the obtained MIS tell us which coefficients in the maximal independence polynomials are non-zeros. Moreover, we can easily reconstruct the found MIS via the indices k (saying which subsets of which size should be taken from the vertices of Q_n).

Table 2. Indices k of MIS in the form $\bigcup_k \binom{[n]}{k}$ along with the MIS sizes for Q_6 and Q_7

$n = 6$	Indices	0, 2, 4, 6	0, 2, 5	0, 3, 5	0, 3, 6	1, 3, 5	1, 3, 6	1, 4, 6	
	Size	32	22	27	22	32	27	22	
$n = 7$	Indices	0, 2, 4, 6	0, 2, 4, 7	0, 2, 5, 7	0, 3, 5, 7	0, 3, 6	1, 3, 5, 7	1, 3, 6	1, 4, 6	1, 4, 7
	Size	64	58	44	58	43	64	49	49	43

4.2 Lattice-Based Results

Let $\mathbb{K}(Q_n) = (2^{[n]}, 2^{[n]}, I)$ be a formal context obtained from the adjacency matrix of a hypercube Q_n as follows: for $g, m \in 2^{[n]}$ $gIm \iff A(Q_n)_{mg} = 1$, where $A(Q_n)$ is the adjacency matrix of Q_n.

Then, the complementary context of $\mathbb{K}(Q_n)$ is $\mathbb{K}^c(Q_n) = (2^{[n]}, 2^{[n]}, (2^{[n]} \times 2^{[n]}) \setminus I)$.

From the comparison of the known number of formal concepts $|\mathbb{K}^c(Q_n)|$ and $mis(Q_n)$ in Table 3, we make the following conjecture.

Table 3. Known $mis(Q_n)$ and $|\mathfrak{B}(\mathbb{K}^c(Q_n))|$ values

n	0	1	2	3	4	5	6
$mis(Q_n)$	1	2	2	6	42	1 670	1 281 402
$|\mathfrak{B}(\mathbb{K}^c(Q_n))|$	1	4	4	36	1 764	2 788 900	1 641 991 085 604

Conjecture 1.
$$mis(Q_n) = \sqrt{|\mathfrak{B}(\mathbb{K}^c(Q_n))|} \ .$$

Property 1. If Conjecture 1 is valid, then the parity of $mis(Q_n)$ coincides with the parity of $|\mathfrak{B}(\mathbb{K}^c(Q_n))|$.

Property 2. If Conjecture 1 is valid, then the number of concepts of $\mathbb{K}^c(Q_n)$ with the non-equal extent and intent of equal sizes is even.

Proof. Since every maximal independent set S is in one-to-one correspondence with (S, S), the formal concept of $\mathbb{K}^c(Q_n)$, $mis(Q_n) = |\mathfrak{M}^+(Q_n)|$, where $\mathfrak{M}^+(Q_n) = \{(S, S) \mid (S, S) \in \mathfrak{B}(\mathbb{K}^c(Q_n))\}$.

Let us denote $mis(Q_n)$ as M_+, then $|\mathbb{K}^c(Q_n)| = M_+^2$. Let us also denote by M_- all the remaining concepts with equally sized extent and intent, not contained in $\mathfrak{M}^+(Q_n)$.

Hence, $M_+^2 = M_+ + M_- + 2l$, where $2l$ counts all the concepts with unequal sizes of extent and intent (it is even due to the existence of concept (B, A) for each (A, B) with $|A| \neq |B|$).

Since, $(M_+ - 1)M_+$ is even, M_- is even.

Table 4. Maximal independence polynomials

n	$I_{max}(Q_n, x)$
0	1
1	$2x^1$
2	$2x^2$
3	$2x^4 + 4x^2$
4	$2x^8 + 16x^5 + 24x^4$
5	$2x^{16} + 32x^{12} + 176x^{10} + 320x^9 + 1140x^8$
6	$2x^{32} + 64x^{27} + 480x^{24} + 1856x^{22} + 8320x^{20}+$ $3840x^{21} + 40320x^{19} + 116320x^{18} + 337920x^{17}+$ $736440x^{16} + 25920x^{15} + 9600x^{14} + 320x^{12}$

Analysing Table 4, with the maximal independence polynomials, we can hypothesise the following statement.

Conjecture 2. 1) For $n > 2$, the second smallest exponent of $I_{max}(Q_n, x)$ is $2^{n-1} - n + 1$, while 2) for $n > 3$, its coefficient is 2^n.

4.3   Spectral-Based Results

Graph Laplacian is used for spectral graph clustering [14], for example, to find communities in social networks. However, we can use this technique for heuristic MIS search for a hypercube as follows:

1. Take the complement of the adjacency matrix for $Q_n = (V, E)$.
2. Compute its Laplacian matrix L.
3. Solve $Lx = \lambda x$.
4. Sort vertices of V by the descending order of x components for every solution. And go to Step 5.
5. Add the sorted vertices one-by-one to a current set S in that order until S is IS.

6. If S is MIS, store it. If there are still non-tested solutions, go to Step 4; stop, otherwise.

The graph Laplacian of $\Gamma = (V, E)$ is $L(\Gamma) = D(\Gamma) - A(\Gamma)$ where $D(\Gamma)$ is the diagonal matrix with $deg(v)$, the degree of vertices $v \in V$, on the main diagonal. The second smallest eigenvalue of this matrix and its eigenvector (also known as Fiedler vector) delivers the minimum value of the $mincut(A, B)$ function over all partitions of V into two parts A and B, i.e. the minimal number of edges that should be deleted from E to make A and B disconnected.

We tested this approach with two more variations of the mincut function (namely, ratio- and normalised cuts) and used eigenvectors for all the larger eigenvalues with the help of scipy.linalg Python module (Fig. 5).

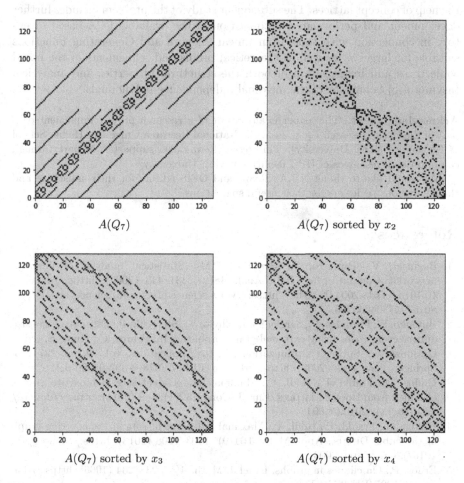

Fig. 5. Adjacency matrix of Q_7 reordered with respect to the smallest eigenvectors of its Laplacian $L(Q_7)$; zero cells are shown in black

Since the tools we used are based on implicitly restarted Arnoldi method to find the eigenvalues and eigenvectors, then we cannot guarantee that all possible sizes of MIS for Q_7 have been found [13]. However, now thanks to this heuristic approach, in addition to the monomials with non-zero coefficients $x^{64}, x^{58}, x^{49}, x^{44}, x^{43}$ obtained in Sect. 4.1, we confirm that the maximal independence polynomial of Q_7 also contains $x^{40}, x^{38}, x^{36}, x^{34}$, and x^{32}.

5 Conclusions

We hope the obtained results both can stimulate the interest of FCA practitioners in the problems of combinatorial enumeration in Graph Theory and attract mathematicians with the discovered facts and posed conjectures obtained with the help of concept lattices. The subsequent study of the problem includes further development and performance optimisation of the available FCA-based inventory in connection with tools from Linear Algebra and Generating Functions suitable for larger n, as well as theoretical proofs and refinements of the posed conjectures and investigation of both the structural properties and analytical (asymptotic) behaviour of the maximal independence polynomials.

Acknowledgement. This paper is an output of a research project implemented as part of the Basic Research Program at the National Research University Higher School of Economics (HSE University). This research was also supported in part through computational resources of HPC facilities at HSE University

We would like to thank N.J.A. Sloane and OEIS editors for their assistance and the anonymous reviewers for their useful suggestions.

References

1. Boginski, V., Butenko, S., Pardalos, P.M.: Statistical analysis of financial networks. Comput. Stat. Data Anal. **48**(2), 431–443 (2005). https://doi.org/10.1016/j.csda.2004.02.004. https://www.sciencedirect.com/science/article/pii/S0167947304000258
2. Butenko, S., Pardalos, P., Sergienko, I., Shylo, V., Stetsyuk, P.: Estimating the size of correcting codes using extremal graph problems. In: Pearce, C., Hunt, E. (eds.) Optimization. Springer Optimization and Its Applications, vol. 32, pp. 227–243. Springer, New York (2009). https://doi.org/10.1007/978-0-387-98096-6_12
3. Duffus, D., Frankl, P., Rödl, V.: Maximal independent sets in bipartite graphs obtained from Boolean lattices. Eur. J. Comb. **32**(1), 1–9 (2011). https://doi.org/10.1016/j.ejc.2010.08.004
4. Duffus, D., Frankl, P., Rödl, V.: Maximal independent sets in the covering graph of the cube. Discret. Appl. Math. **161**(9), 1203–1208 (2013). https://doi.org/10.1016/j.dam.2010.09.003
5. Erdös, P.: On cliques in graphs. Israel J. Math. **4**(4), 233–234 (1966). https://doi.org/10.1007/BF02771637
6. Ganter, B., Obiedkov, S.A.: Conceptual Exploration. Springer, Cham (2016). https://doi.org/10.1007/978-3-662-49291-8

7. Ganter, B., Wille, R.: Formal Concept Analysis - Mathematical Foundations. Springer, Cham (1999). https://doi.org/10.1007/978-3-642-59830-2
8. Hu, H., Mansour, T., Song, C.: On the maximal independence polynomial of certain graph configurations. Rocky Mt. J. Math. **47**(7), 2219–2253 (2017). https://doi.org/10.1216/RMJ-2017-47-7-2219
9. Ignatov, D.I.: Introduction to formal concept analysis and its applications in information retrieval and related fields. In: Braslavski, P., Karpov, N., Worring, M., Volkovich, Y., Ignatov, D.I. (eds.) RuSSIR 2014. CCIS, vol. 505, pp. 42–141. Springer, Cham (2015). https://doi.org/10.1007/978-3-319-25485-2_3
10. Ilinca, L., Kahn, J.: Counting maximal antichains and independent sets. Order **30**(2), 427–435 (2013). https://doi.org/10.1007/s11083-012-9253-5
11. Kahn, J., Park, J.: The number of maximal independent sets in the hamming cube. Comb. **42**(6), 853–880 (2022). https://doi.org/10.1007/s00493-021-4729-9
12. Kuznetsov, S.O.: A fast algorithm for computing all intersections of objects from an arbitrary semilattice. Nauchno-Tekhnicheskaya Informatsiya Seriya 2-Informatsionnye Protsessy i Sistemy (1), 17–20 (1993)
13. Lehoucq, R.B., Sorensen, D.C., Yang, C.: ARPACK Users' Guide: Solution of Large-Scale Eigenvalue Problems with Implicitly Restarted Arnoldi Methods. SIAM, New Delhi (1998)
14. von Luxburg, U.: A tutorial on spectral clustering. Stat. Comput. **17**(4), 395–416 (2007). https://doi.org/10.1007/s11222-007-9033-z
15. Tsukiyama, S., Ide, M., Ariyoshi, H., Shirakawa, I.: A new algorithm for generating all the maximal independent sets. SIAM J. Comput. **6**(3), 505–517 (1977). https://doi.org/10.1137/0206036
16. Weisstein, E.W.: Independence Number. From MathWorld-A Wolfram Web Resource (2023). https://mathworld.wolfram.com/IndependenceNumber.htm
17. Weisstein, E.W.: Independence Polynomial. From MathWorld-A Wolfram Web Resource (2023). https://mathworld.wolfram.com/IndependencePolynomial.html
18. Weisstein, E.W.: Linear Recurrence Equation. From MathWorld-A Wolfram Web Resource (2023). https://mathworld.wolfram.com/LinearRecurrenceEquation.html
19. Xiao, M., Nagamochi, H.: Exact algorithms for maximum independent set. Inf. Comput. **255**, 126–146 (2017). https://doi.org/10.1016/j.ic.2017.06.001

Relational Concept Analysis in Practice: Capitalizing on Data Modeling Using Design Patterns

Agnès Braud[1] , Xavier Dolques[1] , Marianne Huchard[2(✉)] ,
Florence Le Ber[1] , and Pierre Martin[3]

[1] Université de Strasbourg, CNRS, ENGEES, ICube UMR 7537, 67000 Strasbourg,
France
{agnes.braud,dolques}@unistra.fr, florence.leber@engees.unistra.fr
[2] LIRMM, Univ. Montpellier, CNRS, Montpellier, France
marianne.huchard@lirmm.fr
[3] CIRAD, UPR AIDA, 34398 Montpellier, France
pierre.martin@cirad.fr

Abstract. Many applications of Formal Concept Analysis (FCA) and its diverse extensions have been carried out in recent years. Among these extensions, Relational Concept Analysis (RCA) is one approach for addressing knowledge discovery in multi-relational datasets. Applying RCA requires stating a question of interest and encoding the dataset into the input RCA data model, i.e. an Entity-Relationship model with only Boolean attributes in the entity description and unidirectional binary relationships. From the various concrete RCA applications, recurring encoding patterns can be observed, that we aim to capitalize taking software engineering design patterns as a source of inspiration. This capitalization work intends to rationalize and facilitate encoding in future RCA applications. In this paper, we describe an approach for defining such design patterns, and we present two design patterns: "Separate/Gather Views" and "Level Relations".

Keywords: Formal Concept Analysis · Relational Concept Analysis · Design patterns

1 Introduction

Formal Concept Analysis (FCA [15]) has gained importance in knowledge discovery thanks to both theoretical advances and the multiplication of concrete application projects in many domains [31]. Part of these advances are extensions of FCA suitable to handle complex data, going far beyond basic formal contexts. A few of these extensions are dedicated to deal with datasets comprising multiple object categories and multiple relationships between these objects, namely Graph-FCA [12], Relational Concept Analysis (RCA) [17], and the approach described by [22]. Each of these approaches has its own practical application

© The Author(s), under exclusive license to Springer Nature Switzerland AG 2023
D. Dürrschnabel and D. López Rodríguez (Eds.): ICFCA 2023, LNAI 13934, pp. 166–182, 2023.
https://doi.org/10.1007/978-3-031-35949-1_12

and implementation constraints, but has specific qualities for knowledge discovery. E.g. Graph-FCA provides concepts highlighting graph patterns shared by tuples, while RCA provides interconnected concept lattices, one per object category. This paper focuses on RCA.

RCA is based on a simple input data model composed of objects described by Boolean attributes and unidirectional binary relationships between these objects. Practical application raised the issue of encoding a dataset into this formalism, which was more or less easy according to the dataset model structure, such as converting a ternary relation into binary relations [21]. Similar problems arise for encoding object descriptions with particular attribute values (e.g. numerical) into a formal context made of Boolean ones. The latter can be addressed, for example, using scaling approaches [15] or Pattern Structures [14]. To facilitate access of new users to RCA, capitalizing the experience gained in applying RCA in various existing applications is a need.

This paper aims to describe a general approach, to pave the way for the definition of design patterns for RCA application. To this end, we present what such design patterns might look like, and give a few illustrations. Section 2 presents basics of RCA, some typical applications, and our motivation for capitalizing the encoding practices as design patterns. Section 3 outlines the design pattern notion, inspired by its definition in the field of software engineering, and illustrates it through two examples. Section 4 discusses opportunities for developing the approach. We conclude and give a few perspectives of this work in Sect. 5.

2   Background and Motivation

Formal Concept Analysis. FCA is a mathematical framework focusing on the analysis of binary relations using concept lattices. FCA considers as input a formal context $\mathcal{K} = (O, A, I)$ where O is a set of objects, A is a set of attributes and $I \subseteq O \times A$ is the incidence relation. We define the functions $f : \mathcal{P}(O) \to \mathcal{P}(A)$ and $g : \mathcal{P}(A) \to \mathcal{P}(O)$ such that $f(X) = \{y \in A | X \times \{y\} \subseteq I\}$ and $g(Y) = \{x \in O | \{x\} \times Y \subseteq I\}$. The concept lattice $\mathcal{L}$ computed from $\mathcal{K}$ is the set of concepts $\{(X, Y) | X \subseteq O, Y \subseteq A, f(X) = Y \text{ and } g(Y) = X\}$, provided with a partial order relation based on inclusion. X is the concept extent, Y is the concept intent. A concept (X_{sub}, Y_{sub}) is lower in the lattice, i.e. is a sub-concept of a concept (X_{sup}, Y_{sup}) when $X_{sub} \subseteq X_{sup}$.

Relational Concept Analysis. RCA aims at extending FCA to take into account a dataset where objects of several categories are described by attributes and by relations to objects [17]. The dataset is called a Relational Context Family (RCF). An RCF is a $(\mathbf{K}, \mathbf{R})$ pair where: $\mathbf{K} = \{\mathcal{K}_i\}_{i=1,\ldots,n}$ is a set of $\mathcal{K}_i = (O_i, A_i, I_i)$ contexts (i.e. Formal Contexts), and $\mathbf{R} = \{r_j\}_{j=1,\ldots,m}$ is a set of r_j relations (i.e. Relational Contexts) where $r_j \subseteq O_{i_1} \times O_{i_2}$ for some $i_1, i_2 \in \{1, \ldots, n\}$. An example of an RCF composed of two formal contexts introducing pests and crops respectively and one relational context indicating which pest attacks which crop is shown in Table 1. From the RCF, RCA iteratively builds concepts. In a first step, concepts are built for each formal context,

Table 1. Example of a Relational Context Family made of the Formal Contexts `Pest` and `Crop`, and the Relational Context `attacks`.

FC Pest	diptera	beetle	hemiptera	insect	fungi
Contarinia tritici	x			x	
Oulema melanopa		x		x	
Lepidosaphes ulmi			x	x	
Verticillium dalhiae					x

FC Crop	cereal	fruit tree
wheat	x	
barley	x	
apple tree		x
apricot tree		x

RC attacks	wheat	barley	apple tree	apricot tree
Contarinia tritici	x			
Oulema melanopa		x		
Lepidosaphes ulmi			x	
Verticillium dalhiae				x

e.g. concept `C_Crop_15` that groups `wheat` and `barley` for the common attribute `cereal` (right-hand side lattice of Fig. 1). At the next step, the relational context `attacks` is used to form *relational attributes* that express a relation that a pest object may have with a crop concept, such as ∃attacks(C_Crop_15) assigned to `Contarinia triciti` and `Oulema melanopa` because *they attack at least one crop of* `C_Crop_15`. This causes the creation of concept `C_Pest_18`, which would not be there without ∃attacks(C_Crop_15) (left-hand side lattice of Fig. 1). The relational attributes can be formed with different quantifiers (e.g. $\exists\forall$, $\supseteq$, or with percentages). The number of iterations depends on the data model. The same process can be applied to more complex datasets, eventually containing circuits.

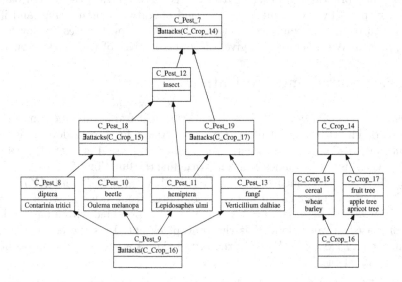

Fig. 1. Lattices obtained from the Relational Context Family of Table 1.

Overview on RCA Applications. RCA was used for analyzing data in various domains. Hydroecological datasets were studied in [11,28], e.g. to correlate physico-chemical parameters and characteristics of taxons living in sample sites. Agro-ecology, and more precisely finding plant-based extracts that can serve as alternative to synthetic pesticides and antimicrobials, was considered in [24]. Using RCA in Information Retrieval (IR) is discussed in [8], and was, for instance, used for querying collections of legal documents connected through cross references [25]. In the field of Semantic Web, it was for instance used to design ontologies [20,32] and extract consistent families of dependent link key candidates in RDF data [4]. RCA was used in industrial information systems to make tools inter-operate, e.g. in the domain of brain activity measurement [35]. It was also applied to identify anomalies in aluminum die casting process description [34]. Finally, RCA was applied in Software Engineering in order to solve different problems: normalize (by factorization) a UML class model [16,19] or a UML use case model [10]; for refactoring [26]; learn model transformation patterns [9]; build directories of Web services or components [5]; structure [7] or analyze [2,18] the variability in software product lines.

Motivation. Applying RCA to a dataset is easy when their data models are compliant. But the model of the dataset is often more complex, and has to be transformed. For instance, it may contain bidirectional or N-ary relations. The relations may have specific semantics, e.g. *is-a*, *instance-of*, and *contains*. In such case, converting an *is-a* relation connecting an entity E_{sub} to another entity E_{super} into an RCA relation conducts to assign attributes of E_{super} to both entities. Identical problems arise when applying FCA to multi-valued attributes. A solution is to use one of the *scaling* schemes [15] that convert non-Boolean attributes into Boolean ones. Mastering these *scaling* schemes is a key for FCA practitioners to select the most appropriate one to meet the modeling objective. The aim of this work is therefore to capitalize on the recurrent data encoding patterns for RCA and thus to increase the efficiency of RCA practitioners in developing applications. In this work, the considered encoding includes values, instances (objects), and data model transformations. The main source of inspiration for this work originated from the domain of Software Engineering, namely the design patterns [13].

3   Design Patterns

This section introduces our proposed description of design patterns (DPs) for RCA (Sect. 3.1) and presents two DPs : *Separate/Gather Views* (Sect. 3.2) and *Level Relations* (Sect. 3.3). The first DP aims to handle (i.e. separates or groups) attributes to facilitate the analysis of a dataset through specific views, and the second one to represent multivalued attributes from the same category discretized into the same set of values using both relational and formal contexts. Both DPs were used in different applications.

3.1   Describing a Design Pattern

Data modeling is a recurrent and fundamental problem in various approaches or domains, including databases, ontologies or software engineering. This last domain is the one where capitalization and sharing of models have been the most developed. The work by Gamma *et al.* [13] is a seminal reference for DP in the domain of Software Engineering. It has been inspired by the work of the architect Christophe Alexander [3]. Gamma *et al.* offers a catalog of DPs for object-oriented designers to transfer knowledge, in particular from experts to beginners, to help them achieve a good design by choosing among several proven alternatives. In the catalog, four essential parts are highlighted to describe a DP: the pattern name, the problem, the solution, and the consequences. The *pattern name* is important as it becomes part of a shared vocabulary. Describing the *problem* involves defining the context and the favorable conditions of the DP application. The *solution* describes the different elements that take part to the design in an abstract and configurable way. The *consequences* include predictable effects of the pattern application and compromises that may have to be done between different qualities of the results. In addition, each DP is described in the catalog using more than ten sections. While some sections are specific to object-oriented design (e.g. classes/objects participants, or sample code), others can be adopted for other domains. In this paper, to remain synthetic, the following five description sections are used:

- **Problem.** The problem is expressed in terms of dataset content and analysis objective. As in [13], this is the most tricky part of the description.
- **Solution.** For RCA, the solution consists in expressing how to formally design and populate a Relational Context Family from the problem description.
- **Example.** This section presents a short example of the pattern application.
- **Known uses.** This section reports existing case studies of the literature where the DP has been applied.
- **Consequences.** This section reviews alternatives and discusses the consequences of the application of the DP relatively to the analysis objective, in particular in terms of usability/readability of the result.

3.2   The Design Pattern *Separate/Gather Views*

Problem. This design pattern applies when:

- The objects identified in the dataset are described either by attributes with domain values of various cardinalities (Case 1), or by Boolean attributes (Case 2). Case 1 can be reduced to Case 2 using a scaling operation [15]. Attributes can be gathered into groups and categories for Case 1 and 2 respectively;
- Each attribute or attribute group of Case 1 or each Boolean attribute category of Case 2 is a coherent *view* on objects;
- It is relevant to analyze objects through the perspective of *a single view* or considering *several views*, first separately and then together.

Solution. The solution, outlined in Fig. 2 (left-hand side), is defined as follows:

- Formal contexts
    - One formal context denoted as FC for the initial objects
        - * FC objects (O) are the initial objects, FC attributes (A) are initial object identifiers, or other description, or none: $FC = (O, A, R), R \subseteq O \times A$

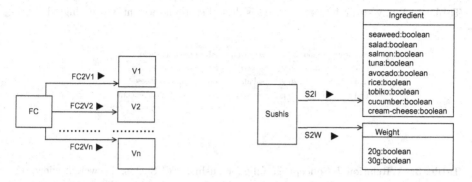

Fig. 2. Schema of the Relational Context Family for the Design Pattern *Separate/Gather Views* (left-hand side). Schema for the sushis example (right-hand side). Both schemas are represented with the UML class diagram notation.

- One formal context denoted as V_i for each view $i, 1 \leq i \leq n$:
  * V_i objects (O_i) are views on the initial objects, V_i attributes (A_i) are Boolean attributes of one view: $V_i = (O_i, A_i, R_i)$, $R_i \subseteq O_i \times A_i$. For each view i, there is a one-to-one mapping between objects of O and their projection in O_i denoted as $proj_i : O \longrightarrow O_i$.
- Relational contexts
  - For each $i, 1 \leq i \leq n$, one relational context $FC2V_i$ connects an object of FC to its corresponding view object in a V_i: $FC2V_i = (O, O_i, r_i)$, $r_i \subseteq O \times O_i$, $r_i = \{(o, proj_i(o)) | o \in O\}$

Example. Table 2 presents the dataset, i.e. sushis described by weight and ingredients, with domain values $\{20g, 30g\}$ and $\{seaweed, ...cream\_cheese\}$ respectively. Applying the DP *Separate/Gather Views* to this dataset consists in identifying the objects and the views. As Sushis are the objects, weight and ingredients are candidates for the views. Figure 2 (right-hand side) graphically outlines the DP application with a UML class model. Class Sushis represents the main formal context; classes Ingredient and Weight represent secondary formal contexts (the views on sushis). UML associations S2I and S2W represent the relational contexts connecting each object to its view object. The values of the attributes Weight and Ingredient lead to Boolean attributes such as seaweed (in Ingredients) or 20g (in Weight). Table 3 shows the resulting RCF: FC is Table Sushis, where there are only objects and no additional description by attributes. V_1 is Table Weight, which associates weight views of sushis (e.g. california salmon Wght) with the corresponding weight (e.g. 20g). V_2 is Table Ingredient, which associates ingredient views of sushis (e.g. california salmon Ing) with the corresponding ingredients (e.g. seaweed). $RC_1 = S2W$ connects a sushi to its weight view; $RC_2 = S2I$ connects a sushi to its ingredient view.

Table 2. A tiny sushi dataset. A sushi is described by its weight and its ingredients.

Sushis	Weight	Ingredients
california salmon	20 g	seaweed, salad, salmon, avocado, rice
california tuna	30 g	seaweed, salad, tuna, avocado, rice
maki cheese	20 g	salad, avocado, rice, cream-cheese
maki tobiko	30 g	seaweed, avocado, rice, tobiko, cucumber

Table 3. A Relational Concept Family for sushis: FC Sushis, V_1 weight view, V_2 ingredient view, $RC_1 = S2W$ connects a sushi to its weight view, $RC_2 = S2I$ connects a sushi to its ingredient view.

FC Sushis
california salmon
california tuna
maki cheese
maki tobiko

V1 Weight	20g	30g
california salmon Wght	x	
california tuna Wght		x
maki cheese Wght	x	
maki tobiko Wght		x

V2 Ingredient	seaweed	salad	salmon	tuna	avocado	rice	tobiko	cucumber	cream-cheese
california salmon Ing	x	x	x		x	x			
california tuna Ing	x	x		x	x	x			
maki cheese Ing		x			x	x			x
maki tobiko Ing	x				x	x	x	x	

RC1 S2W	california salmon Wght	california tuna Wght	maki cheese Wght	maki tobiko Wght
california salmon	x			
california tuna		x		
maki cheese			x	
maki tobiko				x

RC2 S2I	california salmon Ing	california tuna Ing	maki cheese Ing	maki tobiko Ing
california salmon	x			
california tuna		x		
maki cheese			x	
maki tobiko				x

Using this encoding, RCA builds the concept lattices presented in Fig. 3 and 4. Figure 3 presents the views for weight and for ingredients. The weight concept lattice helps analyzing groups of sushis sharing the same weight, i.e. 20g versus 30g. The ingredient concept lattice highlights other sushi groups, e.g. Concept_Ingredient_11 groups the California sushis (separated in the weight view) because of their four shared ingredients (avocado, rice, seaweed, salad). Figure 4 gathers both views and gives combined information, such as the implication rule that a sushi weighed 20g contains salad:
$\exists S2W(\text{Concept\_Weight\_19}) \longrightarrow \exists S2I(\text{Concept\_Ingredient\_12})$.

In this example, the resulting RCA lattice is similar to a FCA lattice. But each view could be more complex, e.g. with hierarchical values or a combination of various types of attributes with consistent semantics. This approach justifies to build a lattice for each view and one lattice gathering all views.

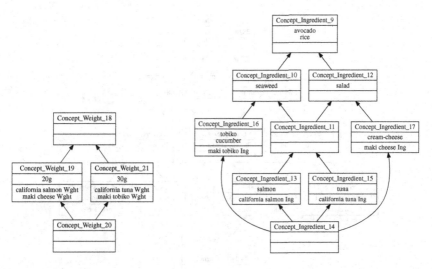

Fig. 3. Views on sushis: according to their weight (left-hand side) and their ingredients (right-hand side).

Known Uses. This DP has been implemented for the analysis of visual accessibility options in operating systems (OS) in [23]. This analysis had different purposes, including making recommendations to OS developers to design a new version, to assist end-users finding an accessibility configuration close to the current configuration when the OS upgrades, or when end-users have to change OS. In this study, the objects are operating systems (OS) and the views cover three visual accessibility options categories (contrast, text, zoom). Separating the views allows to analyze the OS along a single problematic, e.g. to observe commonly shared contrast options, which OS provides more contrast options than another, or which options are never provided together. Gathering the views classifies the OS in a global way, e.g. helping identifying which ones are equivalent on all option categories, how they differ from each other, and how the options of the different categories interact. Moreover, an application has been developed by [18] using this DP to assist *Feature location (FL)* in Software Product Line Engineering.

Consequences. Part of the value of this pattern relies on the relevance of the designed views. It may be more or less complex to determine which set of attributes corresponds to a view, as a single semantics should be associated with this view, e.g. *habitat* versus *food* for animals. Note that partitioning attributes participating to different coherent sets is not relevant, e.g. for animals, the Boolean attribute *aquatic environment* can be an attribute in views *natural habitat* and *growing conditions*. An additional interest of using this RCA DP may occur when considering the pattern variation in which an object can have several views, corresponding for instance to different versions. Different quantifiers can thus be used in the diverse relations, e.g. in the Sushis example,

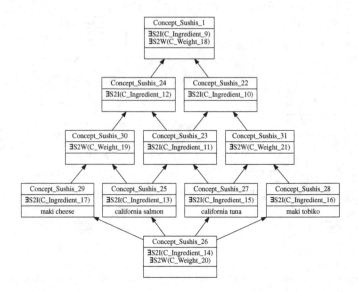

Fig. 4. Sushis concept lattice (gathered views).

$\exists$ on $S2W$ and $\exists\forall$ on $S2I$, to find sushis having versions containing only certain types of ingredients, such as vegan ingredients, considering that this ingredient description was included in V_2 `Ingredient`.

One might wonder whether using FCA on (1) each formal context V_i, and (2) on a simple formal context gathering all the Boolean attributes would be more relevant than applying this pattern, in particular to obtain more readable lattices since relational attributes add some reading complexity. Using the single formal context with all attributes would present these drawbacks: if some attributes of a view are equivalent and correspond to an interesting abstraction (shared by the same object set), then they would be mixed with other attributes of the other views where it will be difficult to observe the equivalence, the implication, or the mutual exclusion between different abstractions coming from different views. In addition, if the description of the attributes is complex, FCA cannot consider this complexity to classify the objects while, in the RCA framework, this complexity can be converted into objects supporting this description. It consists in extending the Relational Context Family in which the whole description is used for classifying the initial objects. To improve the readability of the relational attributes, which is a drawback of RCA, the simplification proposed by [33,34] can be applied: a concept identifier in a relational attribute can be replaced by the intent of the concept at its creation time (possibly recursively). Other authors proposed to deliver the extracted knowledge, not in terms of concepts and relational attributes, but as sentences [29] or as logical formulas in source code macros [18].

3.3   The Design Pattern *Level Relations*

Problem. This design pattern applies when:

- The objects identified in the dataset are described by numerical attributes. The attributes belong to the same semantic category;
- Numerical attributes are discretized into the same set of values;
- Value Levels give interesting information by themselves;
- It is relevant to factorize values to limit the size of the context. Different quantifiers can be applied to the various levels.

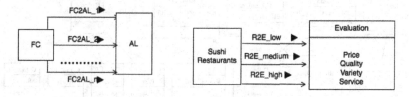

Fig. 5. Schema of the Relational Context Family for the Design Pattern *Level Relations* (left-hand-side). Schema for the sushi restaurant example (right-hand side).

Solution. The solution is outlined in Fig. 5 (left-hand-side) and defined as follows:

- Formal contexts
  - One formal context denoted as FC for the initial objects:
    * FC objects (O) are the initial objects, FC attributes (A) are initial object identifiers, or other description, or none: FC $= (O, A, I)$, $I \subseteq O \times A$
  - One formal context denoted as AL:
    * AL objects (Ω) are the initial attributes, AL attributes (B) are attribute identifiers or categories of attributes: AL $= (\Omega, B, J)$, $J \subseteq \Omega \times B$.
- Relational contexts
  - For $i \in [1, n]$, n being the number of levels, the relational context FC2AL_i connects an object of FC to an attribute in AL if i is the level of this attribute for the object: FC2AL_$i = (O, \Omega, r_i)$, $r_i = \{(o, \omega) | o \in O, \omega \in \Omega, i$ is the value of $\omega(o)$ in the initial dataset$\}$

Example. Table 4 outlines the dataset of the tiny example: sushi restaurants are described by their evaluation on price, food quality, food variety, and service, being multi-valued attributes with the three same values. Figure 5 (right-hand side) graphically outlines the DP application with an UML class model. Class **Sushi Restaurants** represents the main formal context; Class **Evaluation** represents a secondary formal context (the attributes). UML associations **R2E_Low,**

R2E_Medium and R2E_High represent the relational contexts connecting each object to each attribute according to its value level.

Applying the DP *Level Relations* to Table 4 conducted to obtain Table 5. Two objects contexts are defined, i.e. one for the restaurants, the other for the evaluations. Three relational contexts are defined, i.e. one for each evaluation level (low, medium, and high).

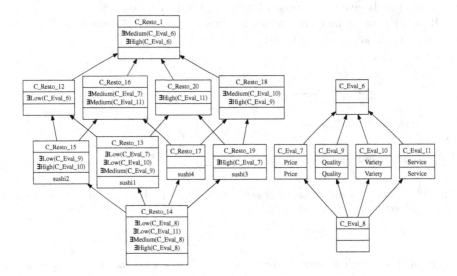

Fig. 6. Sushis restaurant concept lattices. For the sake of space: **Restaurant** has been shorten in **Resto**, **Evaluation** has been shorten in **Eval**, **R2E** has been removed before **Low**, **Medium** and **High**.

Table 4. A tiny sushi restaurant evaluation.

Restaurant	Price	Quality	Variety	Service
sushi1	low	medium	low	high
sushi2	medium	low	high	medium
sushi3	high	high	medium	high
sushi4	medium	high	medium	medium

The lattices obtained from this RCF are shown on Fig. 6. Concepts of the restaurant lattice (left on Fig. 6) have attributes pointing to evaluation concepts (right on Fig. 6). For instance, concept **C_Resto_16** gathers restaurants that have a medium evaluation on Price and on Service. Concept **C_Resto_1** gathers all restaurants, showing that all have both a medium and a high evaluation on some criteria.

Table 5. A Relational Context Family for sushis restaurants: *FC* Restaurant, *AL* Evaluation. Identifiers of objects are not shown for a sake of space. R_1=R2E-Low connects a sushi restaurant to its low evaluations, R_2=R2E-Medium connects a sushi restaurant to its medium evaluations, R_3=R2E-High connects a sushi restaurant to its high evaluations.

Restaurants
sushi1
sushi2
sushi3
sushi4

Evaluation
Price
Quality
Variety
Service

R2E-Low	Price	Quality	Variety	Service
sushi1	x		x	
sushi2		x		
sushi3				
sushi2				

R2E-Medium	Price	Quality	Variety	Service
sushi1		x		
sushi2	x			x
sushi3			x	
sushi2	x		x	x

R2E-High	Price	Quality	Variety	Service
sushi1				x
sushi2			x	
sushi3	x	x		x
sushi2		x		

Known Uses. DP *Level Relations* has been applied for modeling a water dataset, where stream sites are described by physico-chemical parameters (pH, nitrates, phosphates, etc.) and fauna information [11]. Data are modeled with several relations to represent the different value levels of physico-chemical parameters (five level relations), macro-invertebrate populations (three level relations), and characteristics (life traits) of the macro-invertebrates (six level relations). Finally the RCF contains four formal contexts (stream sites, physico-chemical parameters, life traits and macro-invertebrates) and 14 relational contexts.

Moreover, this DP was also used by [27] to analyze a dataset on hydrosystem restoration projects. In this work, river sites are characterized by temporal heterogeneous information, including measures of physico-chemical parameters, biological indicators, and the composition of the surrounding land use. Biological indicators are described with 5 quality values. Physico-chemical parameters are discretized into 5 values. Land use is assessed within two increasing buffers. Part of different types of land use is summarized into three values, low, medium and high. The river sites belong to river segments where restoration operations have been undertaken. Finally the RCF contains 6 formal contexts and 21 relational contexts.

Consequences. The lattice on sushi restaurants (left on Fig. 6, called lattice R in the following) obtained with RCA based on the DP *Level Relations* applied on the dataset described in Table 4 can be compared with a lattice obtained with a simple scaling of the multi-valued attributes into Boolean ones. This second lattice, called lattice B in the following, is shown in Fig. 7. Concepts introducing the objects (`restaurant`) are similar in both lattices. More general concepts are different: lattice R has one more concept, i.e. `C_Resto_12`, which groups restaurants having a low evaluation on some criteria. In such case, RCA produces relational attributes in which only the level ($\exists$ *medium* / *high*) is represented and not the value (which attribute). Such attributes can be interesting for experts.

Furthermore the tops of the lattices are different: it is empty in lattice B while it contains two relational attributes in lattice R. DP *Level Relations* is thus relevant when values provide an interesting information.

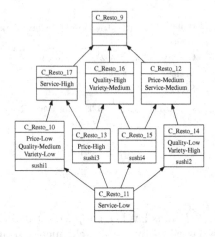

Fig. 7. Lattice obtained from Table 4 with a scaling of multi-valued attributes.

4   Perspectives

Section 3 presented two DPs, that were used in concrete applications, focusing on various aspects of relational data, as can be considered in the RCA framework.

Several perspectives are offered by this work. Two of them are discussed in this section, i.e. identifying additional DPs that can be useful to help RCA users and revisiting existing applications.

New Opportunities for Defining DPs. Additional DPs were identified along the existing applications. A few examples are introduced hereafter. Some data models included a specialization/generalization (*is-a*) relationship. Analyzing the associated dataset required to flatten the hierarchical descriptions [10,16,19], to discover abstractions previously hidden. This new DP could be named **Collapse Specialization**. The DP **Reify Relation** may consist in introducing a formal context in the RCF for describing the tuples of a relation. This approach was used by [21] to derive a binary representation of an N-ary relation, without loosing information. The link reification, as realized in [9] and [4], could be considered through the same perspective. The DP, dual to *Separate/Gather Views*, is **Separate/Gather Objects**, in which objects of a same category can be separated into several subsets to analyze each subset apart or all subsets as a whole. This approach was considered by [16] to normalize a UML class model to detect candidate attribute generalizations on diverse criteria, e.g. name, substring in the name, synonyms, type, default value. The DP **Instances2Model** may address objects with multi-valued attributes to extract a schema, as it has been done from instance descriptions in [9]. It consists in simplifying the description considering that an object has a value for an attribute (but not the value). Introducing virtual objects representing a **Query** has been proposed by authors in the context of FCA. This can be extended to the context of RCA, such as in [6], introducing several virtual objects in formal contexts and virtual links in relational contexts. This approach has been adopted for solving the problem of replacing a failing web service in a web service workflow [5]. Finally, sequences can be modeled into RCA input with a transitive relation 'precedes' or 'succeeds', according to the DP **Temporal** [28]. Other transitive relations (e.g. part-of/includes, or down/upstream-of [27]) may also correspond to this DP.

Revisiting Applications with the DPs. The DP *Separate/Gather Views* presented in Sect. 3 could be applied to revisit and improve a previous work on component catalog FCA-based building [1], in which software components were described with provided/required interfaces, each interface containing a set of services. In this catalog, the description was made using a single formal context. The catalog then organized the software components with this description and exposed possible substitution between components from the two viewpoints as a whole (provided as well as required services). This hardly helps the lattice exploration considering only one viewpoint, e.g. a user may want to search a component with provided services, and consider in a second step the required services, that other components could provide. Another potential use of the DP *Separate/Gather Views* could be to consider positive versus negative description of objects, a question that has been addressed in [30], to complete their approach using an additional point of view.

5 Conclusion

This paper presented a general approach that aims to capitalize lessons learned in encoding datasets in the RCA input format. Two DPs were described to illustrate the approach. These DPs have been extracted from existing RCA applications, and could be applied to improve or complete the analysis in other applications. Additional DPs to formalize were also reported. Short-term future work includes continuing the DPs formalization and defining a data modeling process involving the DPs. In the longer term, a new area of research could be extending this approach to other FCA extensions, such as Pattern Structures, Graph-FCA, and Polyadic FCA. Using DPs could be a gateway to use several FCA extensions in synergy, as they could improve their implementation in case studies.

Acknowledgements. This work was supported by the ANR SmartFCA project, Grant ANR-21-CE23-0023 of the French National Research Agency.

References

1. Aboud, N., et al.: Building hierarchical component directories. J. Object Technol. **18**(1), 2:1–37 (2019)

2. Al-Msie'deen, R., Seriai, A., Huchard, M., Urtado, C., Vauttier, S.: Documenting the mined feature implementations from the object-oriented source code of a collection of software product variants. In: 6th International Conference on Software Engineering and Knowledge Engineering (SEKE), pp. 138–143 (2014)
3. Alexander, C.: A Pattern Language: Towns, Buildings, Construction. Oxford University Press, Oxford (1977)
4. Atencia, M., David, J., Euzenat, J., Napoli, A., Vizzini, J.: Link key candidate extraction with relational concept analysis. Discret. Appl. Math. **273**, 2–20 (2020)
5. Azmeh, Z., Driss, M., Hamoui, F., Huchard, M., Moha, N., Tibermacine, C.: Selection of composable web services driven by user requirements. In: IEEE International Conference on Web Services (ICWS), pp. 395–402. IEEE Computer Society (2011)
6. Azmeh, Z., Huchard, M., Napoli, A., Hacene, M.R., Valtchev, P.: Querying relational concept lattices. In: 8th International Conference on Concept Lattices and Their Applications (CLA). Proceedings of CEUR Workshop, vol. 959, pp. 377–392 (2011)
7. Carbonnel, J., Huchard, M., Nebut, C.: Modelling equivalence classes of feature models with concept lattices to assist their extraction from product descriptions. J. Syst. Softw. **152**, 1–23 (2019)
8. Codocedo, V., Napoli, A.: Formal concept analysis and information retrieval – a survey. In: Baixeries, J., Sacarea, C., Ojeda-Aciego, M. (eds.) ICFCA 2015. LNCS (LNAI), vol. 9113, pp. 61–77. Springer, Cham (2015). https://doi.org/10.1007/978-3-319-19545-2_4
9. Dolques, X., Huchard, M., Nebut, C., Reitz, P.: Learning transformation rules from transformation examples: an approach based on relational concept analysis. In: Workshops on Proceedings of the 14th IEEE International Enterprise Distributed Object Computing Conference (EDOCW), pp. 27–32 (2010)
10. Dolques, X., Huchard, M., Nebut, C., Reitz, P.: Fixing generalization defects in UML use case diagrams. Fundam. Inf. **115**(4), 327–356 (2012)
11. Dolques, X., Le Ber, F., Huchard, M., Grac, C.: Performance-friendly rule extraction in large water data-sets with AOC posets and relational concept analysis. Int. J. Gen Syst **45**(2), 187–210 (2016)
12. Ferré, S., Cellier, P.: Graph-FCA: an extension of formal concept analysis to knowledge graphs. Discrete Appl. Math. **273**, 81–102 (2020)
13. Gamma, E., Helm, R., Johnson, R., Vlissides, J.: Design Patterns: Elements of Reusable Object-oriented Software. Addison-Wesley Longman, Boston (1995)
14. Ganter, B., Kuznetsov, S.O.: Pattern structures and their projections. In: Delugach, H.S., Stumme, G. (eds.) ICCS-ConceptStruct 2001. LNCS (LNAI), vol. 2120, pp. 129–142. Springer, Heidelberg (2001). https://doi.org/10.1007/3-540-44583-8_10
15. Ganter, B., Wille, R.: Formal Concept Analysis - Mathematical Foundations. Springer, Cham (1999)
16. Guédi, A.O., Miralles, A., Huchard, M., Nebut, C.: A practical application of relational concept analysis to class model factorization: lessons learned from a thematic information system. In: 10th International Conference on Concept Lattices and Their Applications (CLA). CEUR Workshop Proceedings, vol. 1062, pp. 9–20 (2013)
17. Hacene, M.R., Huchard, M., Napoli, A., Valtchev, P.: Relational concept analysis: mining concept lattices from multi-relational data. Ann. Math. Artif. Intell. **67**(1), 81–108 (2013)

18. Hlad, N., Lemoine, B., Huchard, M., Seriai, A.: Leveraging relational concept analysis for automated feature location in software product lines. In: The ACM SIGPLAN International Conference on Generative Programming: Concepts & Experiences (GPCE), Chicago, IL, USA, pp. 170–183. ACM (2021)

19. Huchard, M., Hacene, M.R., Roume, C., Valtchev, P.: Relational concept discovery in structured datasets. Ann. Math. Artif. Intell. **49**(1–4), 39–76 (2007)

20. Kasri, S., Benchikha, F.: Refactoring ontologies using design patterns and relational concepts analysis to integrate views: the case of tourism. Int. J. Metadata Semant. Ontol. **11**(4), 243–263 (2016)

21. Keip, P., Ferré, S., Gutierrez, A., Huchard, M., Silvie, P., Martin, P.: Practical comparison of FCA extensions to model indeterminate value of ternary data. In: 15th International Conference on Concept Lattices and Their Applications (CLA). CEUR Workshop Proceedings, vol. 2668, pp. 197–208 (2020)

22. Kötters, J., Eklund, P.W.: Conjunctive query pattern structures: a relational database model for formal concept analysis. Discrete Appl. Math. **273**, 144–171 (2020)

23. Kouhoué, A.W., Bonavero, Y., Bouétou, T.B., Huchard, M.: Exploring variability of visual accessibility options in operating systems. Fut. Internet **13**(9), 230 (2021)

24. Mahrach, L., et al.: Combining implications and conceptual analysis to learn from a pesticidal plant knowledge base. In: Braun, T., Gehrke, M., Hanika, T., Hernandez, N. (eds.) ICCS 2021. LNCS (LNAI), vol. 12879, pp. 57–72. Springer, Cham (2021). https://doi.org/10.1007/978-3-030-86982-3_5

25. Mimouni, N., Fernández, M., Nazarenko, A., Bourcier, D., Salotti, S.: A relational approach for information retrieval on XML legal sources. In: International Conference on Artificial Intelligence and Law (ICAIL), pp. 212–216. ACM (2013)

26. Moha, N., Rouane Hacene, A.M., Valtchev, P., Guéhéneuc, Y.-G.: Refactorings of design defects using relational concept analysis. In: Medina, R., Obiedkov, S. (eds.) ICFCA 2008. LNCS (LNAI), vol. 4933, pp. 289–304. Springer, Heidelberg (2008). https://doi.org/10.1007/978-3-540-78137-0_21

27. Nica, C., Braud, A., Le Ber, F.: Exploring heterogeneous sequential data on river networks with relational concept analysis. In: Chapman, P., Endres, D., Pernelle, N. (eds.) ICCS 2018. LNCS (LNAI), vol. 10872, pp. 152–166. Springer, Cham (2018). https://doi.org/10.1007/978-3-319-91379-7_12

28. Nica, C., Braud, A., Le Ber, F.: RCA-SEQ: an original approach for enhancing the analysis of sequential data based on hierarchies of multilevel closed partially-ordered patterns. Discrete Appl. Math. **273**, 232–251 (2020)

29. Ouzerdine, A., Braud, A., Dolques, X., Huchard, M., Le Ber, F.: Adjusting the exploration flow in relational concept analysis - an experience on a watercourse quality dataset. In: Jaziri, R., Martin, A., Rousset, M.C., Boudjeloud-Assala, L., Guillet, F. (eds.) Advances in Knowledge Discovery and Management, Studies in Computational Intelligence, vol. 1004, pp. 175–198. Springer, Cham (2019)

30. Pérez-Gámez, F., Cordero, P., Enciso, M., López-Rodríguez, D., Mora, Á.: Computing the mixed concept lattice. In: Davide, C., et al., (eds.) Information Processing and Management of Uncertainty in Knowledge-Based Systems (IPMU), IPMU 2022, vol. 1601 pp. 87–99. Springer, Cham (2022). https://doi.org/10.1007/978-3-031-08971-8_8

31. Poelmans, J., Ignatov, D.I., Kuznetsov, S.O., Dedene, G.: Formal concept analysis in knowledge processing: a survey on applications. Expert Syst. Appl. **40**(16), 6538–6560 (2013)

32. Rouane Hacene, A.M., Napoli, A., Valtchev, P., Toussaint, Y., Bendaoud, R.: Ontology learning from text using relational concept analysis. In: International MCETECH Conference on e-Technologies (2008)
33. Wajnberg, M.: Analyse relationnelle de concepts : une méthode polyvalente pour l'extraction de connaissance. (Relational concept analysis: a polyvalent tool for knowledge extraction). Ph.D. thesis, Univ. du Québec à Montréal (2020)
34. Wajnberg, M., Valtchev, P., Lezoche, M., Massé, A.B., Panetto, H.: Concept analysis-based association mining from linked data: a case in industrial decision making. In: Joint Ontology Workshops 2019 Episode V: The Styrian Autumn of Ontology. CEUR Workshop Proceedings, vol. 2518. CEUR-WS.org (2019)
35. Wajnberg, M., Lezoche, M., Blondin-Massé, A., Valchev, P., Panetto, H., Tyvaert, L.: Semantic interoperability of large systems through a formal method: relational concept analysis. IFAC-PapersOnLine **51**(11), 1397–1402 (2018)

Representing Concept Lattices with Euler Diagrams

Uta Priss[✉][iD]

Fakultät Informatik, Ostfalia University, Wolfenbüttel, Germany
u.priss@ostfalia.de
http://www.upriss.org.uk/

Abstract. The Hasse diagrams of Formal Concept Analysis (FCA) concept lattices have the disadvantages that users need to be trained in reading the diagrams and diagrams of larger lattices tend to be too cluttered to be comprehensible. This paper therefore discusses how to reduce lattices and then represent them with a specific type of Euler diagram instead of Hasse diagrams. A semi-automated process of reducing concept lattices is described and supported by algorithms.

1 Introduction

Formal Concept Analysis (Ganter and Wille, 1999) provides a mathematical method for data analysis which includes graphical representations in form of Hasse[1] diagrams of concept lattices[2]. Unfortunately diagrams for concept lattices with more than a small number of concepts contain many edges rendering them difficult to visually parse. Without being trained in reading diagrams, people often misread Hasse diagrams, for example, by mistaking the edges to be vectors[3]. Small Euler diagrams, however, can usually be read by people without any significant amount of training[4]. But Euler diagrams require a small size and careful layout algorithms in order to be comprehensible. Therefore two questions arise: first, how can formal contexts be reduced or partitioned so that their diagrams are sufficiently small to be easier to represent and, second, in order to avoid training users in reading diagrams, how can concept lattices be represented as Euler diagrams?

We are arguing in this paper that an optimal layout of Hasse or Euler diagrams cannot be provided via a purely deterministic algorithm because it may also depend on the content of the data. For example, some objects or attributes may be similar, belong to a shared domain or enhance or contradict each other and should be represented in a manner that reflects such relationships. These other relationships amongst objects

---

[1] The diagrams should not be named after Hasse because he did not invent them, but the name is widely established in the literature.

[2] An introduction to FCA is not included in this paper. Standard FCA terminology is used: *intension, extension*, a *cross* table for the formal context, and so on. An *object concept* is the lowest concept which has the object in its extension. *Attribute concepts* are defined dually.

[3] Based on the author's personal teaching experience with introductory mathematics courses.

[4] Again based on the author's teaching experience. "Small" in this paper means < 20 concepts.

D. Dürrschnabel and D. López Rodríguez (Eds.): ICFCA 2023, LNAI 13934, pp. 183–197, 2023.
https://doi.org/10.1007/978-3-031-35949-1_13

or attributes can be considered *background knowledge* which might change with each application and need not necessarily be mathematically precise. Therefore, instead of a single algorithm that produces optimal layouts, a semi-automated process is favourable that uses heuristics and interacts with users by asking them questions about the data and by giving them choices between layout options. Such a process is similar to "conceptual exploration" (Ganter and Obiedkov 2016) but conceptual exploration tends to modify the data contained in formal contexts whereas this paper assumes that only the diagrams are modified but not the original data.

Most existing research papers that discuss FCA diagram layouts or data reduction are not relevant for this paper because their lattices are still larger than in this paper, use fuzzy or probabilistic reduction methods or focus on navigation through a lattice (e.g. Vogt and Wille (1995), Cole, Eklund and Stumme (2003) or Alam, Le and Napoli (2016)). There is a large body of research on visualisations with Euler diagrams (e.g. Chapman et al. 2014), but not relating to concept lattices. The aim of this paper is to improve visualisations of conceptual hierarchies that are small, non-fuzzy but with background knowledge, for example visualisations of relationships in lexical databases (Priss and Old 2010) and of conceptual representations used for teaching. A goal is to improve existing software[5] by providing fairly simple generic algorithms for operations as outlined in the Appendix. Mathematically, such algorithms need not be complicated. Complexity considerations are not relevant for small data sets and not discussed in this paper. A challenge, however, is to respect user preferences and preserve implicit information within the data according to background knowledge. Furthermore, generating Euler diagrams is more challenging than generating Hasse diagrams as explained in the next section.

Digital tools for mathematics education often rely on "multiple linked representations (MLR)" which connect different visualisations so that changes in one part affect changes in other parts (Ainsworth 1999). In the case of FCA, an MLR display can show a formal context, its lattice and its implication base simultaneously. Changes in the context will then, for example, instantly change the lattice and the implication base[6] Pedagogically, MLRs are useful because according to educational research, mathematical thinking often involves simultaneously imagined representations and fast shifting between them. For example, Venn and Euler diagrams are suitable means for teaching introductory set theory. But the fact that the empty set is a subset of every other set cannot be shown with a Venn or Euler diagram. It can only be concluded from the definition of "subset" by logical reasoning. Thus students need to learn that some knowledge can be represented diagrammatically (such as the subset relationship in Euler diagrams) whereas other knowledge must be represented textually using logical formulas.

Applying the idea of MLRs to FCA, one can argue that it is not necessary to represent all information contained in a formal context as a diagram of a concept lattice but instead that some information can be presented diagrammatically and some textually using implications or other logical statements. Therefore a strategy for reducing concept lattices proposed in this paper is to offload some information from diagrams into

---

[5] https://github.com/upriss/educaJS.

[6] An example of such an MLR of context and lattice can be found at https://upriss.github.io/educaJS/binaryRelations/fcaExample.html.

a textual representation. In some cases "a picture is worth a 1000 words". But in other cases, textual representations are simpler. For example, compared to drawing a Boolean lattice, a statement that all attributes from a set occur in any possible combination may be simpler. The decision as to which information should be presented diagrammatically or textually is often not based on mathematics but instead on background knowledge.

The next section introduces the specific type of Euler diagram that is suggested in this paper. Section 3 compiles a list of reduction methods that are applicable to non-fuzzy, non-probabilistic data. Section 4 demonstrates how in particular a reduction method that utilises concept extensions for partitioning objects is promising for generating Euler diagrams. Section 5 provides a short conclusion. An Appendix contains algorithms for the reduction methods.

2   Rounded Rectangular Euler Diagrams

This section argues that rounded rectangular Euler diagrams are ideally suited for representing small concept lattices and other set inclusion hierarchies. Euler diagrams are a form of graphical representation of set theory that is similar to Venn diagrams but leaves off any zones that are known to be empty (Fig. 1). Most authors define the notion of "Euler diagram" in a semi-formal manner (Rodgers 2014) because a formal definition that captures all aspects of an Euler diagram is complex. Furthermore, mainly a certain subset of *well-formed* diagrams is of interest but so far no algebraic characterisation of that subset has been found. Euler diagrams consist of closed curves with labels representing sets. The smallest undivided areas in an Euler diagram are called *minimal regions* (Rodgers 2014). Regions are defined as sets (or unions) of minimal regions. *Zones* are maximal regions that are within a set of curves and outwith the remaining curves. In this paper, Euler diagrams are required to be well-formed in the sense that at most two curves intersect in any point and that each zone must not be split into several regions. Thus it is not necessary to distinguish between zones and minimal regions. The left half of Fig. 1 shows a well-formed Euler diagram with 6 zones (including the outer zone).

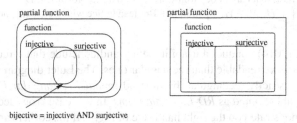

Fig. 1. Rectangular Euler diagrams with and without rounded corners and distortion

It is generally accepted that well-formed diagrams are easier to read than non-well-formed diagrams (Rodgers 2014) but not all subsets of a powerset can be represented as a well-formed diagram. In order to represent more, *shading* can be applied to zones that are meant to be empty. With shading any subset of a powerset can be drawn as a well-formed Venn or Euler diagram. But shading several non-adjacent zones also renders diagrams difficult to read and thus should be used sparingly. There is some indication that being a lattice, or at least complete with respect to intersections, increases the probability of being representable as a well-formed Euler diagram but that is neither a necessary nor a sufficient condition (Priss 2020). Tree-like set hierarchies can always be represented as well-formed Euler diagrams. For lattices, it helps if one is allowed to omit the bottom concept, thus rendering the structure slightly more tree-like.

A further common condition for being well-formed is that curves must not be concurrent which means disallowing curves to meet in more than one adjacent point. This condition is actually dropped in this paper because it allows more subsets of a powerset to be represented. For example, a set of 3 elements has a powerset of 8 elements which has $2^8 = 256$ possible different subsets. If one considers only those where the outer zone is not empty (i.e. 128 subsets), only about 35 of these can be drawn as well-formed Euler diagrams without shading if each set is represented as a circle. If sets are represented as rectangles and curves are allowed to be concurrent, then we counted about 100 subsets that are representable without shading. The left and middle images in Fig. 2 show two examples that are representable in the manner suggested in this paper, but not representable using circles without shading. Boolean lattices with more than 4 sets cannot be represented as rectangular Euler diagrams, but are also difficult to read as Hasse diagrams. A fifth set can be added as a concave curve, as shown with the dashed line in Fig. 2 (right image). Because not all formal contexts with more than 4 elements are Boolean, it is still possible to represent some concept lattices with more than 4 attributes with rectangular Euler diagrams.

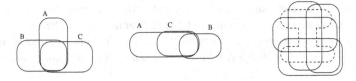

Fig. 2. Euler diagrams that cannot be "well-drawn", two dimensional $\{A, B, C, ABC\}$ (left) and one dimensional $\{A, B, AC, ABC\}$ (middle). Boolean lattice with 5 elements on the right (three dimensional).

Blake et al. (2014) conduct a usability study and conclude that circular Euler diagrams are much more readable than rectangular ones. The Euler diagrams in this paper, however, are not strictly rectangular, but instead *rounded rectangular Euler diagrams with distortion*, abbreviated as *RD-Euler diagrams*. In this case "rounded" refers to the corners. As demonstrated on the right hand side of Fig. 1, if the corners are not rounded, then all sets and their intersections have the same shape and are difficult to distinguish. In circular diagrams only the sets are circles, whereas the intersections are composed of circle segments. Using RD diagrams (i.e. with purposeful imperfections) has the effect of changing the shape of the intersections and rendering them easier to discern. An alternative would be to use different colours for different sets but that potentially interferes with shading. Because of the distortion it is possible to see the shape of each set in Fig. 2. The small intersections between the concurrent edges are meant to not exist or in other words be empty. It is assumed that users will consider them drawing errors and ignore them (but a usability study should determine which size and shape of distortion is still acceptable to users and at what point shading is required).

Figure 2 also shows that in some cases the sets and their intersections can be arranged in a linear sequence as *one dimensional diagrams*. Petersen (2010) describes the conditions under which concept lattices can be linearised in that manner. Chapman et al. (2014) observe that one dimensional linear diagrams (containing lines instead of curves) have a higher usability for certain tasks than Euler diagrams because users read them faster and more accurately even though instead of shading such diagrams use repetition of attributes. Two dimensional diagrams can be embedded into direct products of one dimensional diagrams. Thus for small formal contexts a simple algorithm consists of repeatedly splitting the sets of attributes in half and then checking for linearity in order to compute a representation as an RD-Euler diagram. We are arguing that two dimensional diagrams are not significantly more complicated than one dimensional ones because only two directions (horizontal and vertical) need to be observed and should thus have similar advantages as observed by Chapman et al. (2014). Three dimensional diagrams (Fig. 2 on the right) are difficult to read unless the sets of the third dimension are small and convex.

Apart from being able to express more subsets of a powerset than circular Euler diagrams, another advantage is that RD-Euler diagrams are easier to shade and label when drawn with software. In order to shade intersections one needs to compute their exact geometrical shape which is easier for rectangles than for circles. Furthermore it is more difficult to attach labels to circles. For rectangles, labels can be written adjacent to the edge of a curve either exactly horizontally or exactly rotated by 90°. Intersections cannot be labelled in that manner and do require a line or arrow as in the left half of Fig. 1 for either circular or RD-Euler diagrams.

3   Reducing Concept Lattices

This section provides a brief summary of methods for reducing concept lattices by offloading information from diagrams and instead storing it as logical expressions. The resulting RD-Euler diagrams then focus on information that is easier to represent graphically whereas the expressions focus on information that is more suited to textual representation. As mentioned in the introduction, the topic of reducing lattices has been discussed in the literature but usually with a focus on special properties or using fuzzy/probabilistic methods. The methods discussed in this section are more general, simple and mostly well-known. But the practical, joint use of diagrams and textual information has not yet received much attention. The assumptions about the data of a given formal context for this paper are:

crisp, non-fuzzy data: probability or other approximations are not relevant for this paper.
background knowledge may be available providing further relations, groupings, domains and
   so on for the objects and attributes.
three valued logic applies to most formal contexts because a missing cross can either mean that
   an attribute does not apply or that it is unknown whether it applies.

negatable attributes use binary logic in that objects either definitely do or do not have such attributes according to background knowledge.

supplemental concepts are concepts that can never be object concepts according to background knowledge even if more objects are added to a formal context (Priss 2020).

A formal context can be split into parts that are implications, diagrams and cross tables as long as it is clear how the original context can be recomputed from the different parts. A step-wise reduction should start by storing the original set of implications, because reduction can change implications. At any reduction step, if implications are lost, they must be extracted and stored as well. As a final step, only the implications that are not visible from the diagrams should be displayed textually. How to structure implications so that they are most effective for human users is not trivial and still an open research question that has been discussed elsewhere (Ganter 2019). Using general logical expressions instead of implications complicates matters but only if logical properties are to be computed, not if information is just displayed. Furthermore, as discussed in the introduction, it is assumed that the methods described here are executed in a semi-automated manner. Users will be asked questions about the data and then be presented with diagrams and texts using different reduction methods. Users can then choose which displays they prefer and whether to stop or try further methods.

The following listing summarises reduction methods, many of which are well known in some form, but according to our knowledge the approach of offloading information from each reduction step into a textual expression has not yet been studied much. The first four methods are discussed by Priss (2021). They can be easily implemented via an algorithm that searches through a formal context as shown in the Appendix. In our opinion the last method is particularly interesting and has not yet received as much attention in the literature. As a rule, contexts with fewer crosses result in less complicated diagrams. Thus a guiding strategy is to remove attributes and crosses or to split contexts.

SYNONYM-reduction: if several attributes have the same extensions then all but one can be removed. Expressions should be added using ":=".

AND-reduction: an attribute whose attribute concept is a meet of other concepts can be removed. An expression using "AND" should be added (e.g. "bijective" in Fig. 1).

OR-reduction: an attribute whose attribute concept is a supplemental concept can be removed if its extension is a union of extensions of other attribute concepts. An expression using "OR" should be added. Caution: if several OR-reducible attributes exist, they cannot be removed simultaneously because removal of one may affect the others.

NOT-reduction: if an attribute is a negation of another attribute and both are negatable, the one with more crosses can be removed. An expression using "NOT" should be added.

NEGATION-reduction: if an attribute is negatable and its column in a context is filled more than half with crosses, then it can be replaced by its negation. An expression using "NOT" should be added.

FACTORISATION-reduction: if a context can be factorised using binary matrix multiplication so that the set of factors is smaller than the original set of attributes or smaller than a chosen value, then the set of factors can replace the set of original attributes (Belohlavek and Vychodil 2010). An expression using "AND" should be stored for each factor. Caution: factors cut the lattice vertically in half. Thus the height of the lattice is reduced but the horizontal structure at the level of the cut remains the same. Also, background knowledge may determine whether or not factors are a reasonable choice for a domain.

HORIZONTAL SPLIT: applicable if removing the top and the bottom concepts (which must both be supplemental) partitions the Hasse diagram into several separate graphs. The only implications that involve attributes from different parts use the bottom concept and may not even be of interest according to background knowledge. If the different parts have nothing in common, then no expressions need to be stored.

LOWER HORIZONTAL SPLIT: if removing the bottom concept and some supplemental concepts other than the top concept partitions the Hasse diagram into several separate graphs so that all object concepts are below the removed concepts. (This is different from factorisation because only object concepts matter for this type of splitting.)

PARTITIONING attributes: this is commonly used in FCA, for example, for nested line diagrams (Ganter and Wille, 1999). Caution: it changes the set of implications and there is no easy manner in which to store the lost information as expressions.

PARTITIONING objects: no implications are lost but each partition may have additional implications which do not hold in the original context.

CONCEPTUAL-PARTITIONING: same as the previous one but the set of objects is partitioned so that each partition is an extension of a concept or an extension of a concept of a negated attribute. The additional implications that are generated can be expressed using quantification over extensions: "FOR ALL ... that belong to extension ... : ...".

AND-reduction does not change a concept lattice. The methods of OR-and NOT-reduction and horizontal split are very effective at reducing a diagram, but the conditions for applying these are not met by many formal contexts. Factorisation depends on background knowledge and might not reduce the complexity of a diagram significantly. Negation reduction is very promising but requires a binary logic for its attribute. Therefore, partitioning appears to be the most interesting reduction method. As far as we are aware, the Darmstadt FCA group around Rudolf Wille favoured partitioning attributes over partitioning objects. As explained below, at least with respect to implications, partitioning objects is a much more promising choice. In our opinion, while nested line diagrams have proven to be useful for some applications, the diagrams are not easy to visually comprehend as a whole. If users zoom into one part of a nested diagram, they need to memorise where they are and what the relevant outer attributes are.

For conceptual partitioning, the set of objects is partitioned so that each partition can be expressed using statements involving attributes. For example, if the extensions of attributes a and b partition the set of objects, then the two partitions can be referred to as "objects having a" and "objects having b". If attributes are negatable, then one can also partition into "objects having a" and "objects not having a" or into "objects having a", "objects having b" and "objects having neither a nor b". Most likely negation should be used sparingly and not nested in order to avoid creating very complicated statements. For the same reason, not too many partitions should be created and it is ideal if the attributes chosen provide some obvious categorisation according to background knowledge (for example, "people in their 20s", "people in their 30s", and so on). Horizontal split is a type of conceptual partitioning because the extensions of the immediate lower neighbours of the top concept partition the set of objects. Lower horizontal split is a type of conceptual partitioning if the objects of each part exactly correspond to an extension of a concept. Figure 3 in the next section shows an example.

With regard to evaluating the effectiveness of reduction methods, a few simple measures can be defined:

M_{cpt}: total number of concepts in a lattice
M_{obj}, M_{att}, M_{oa}: numbers of object and attribute concepts and concepts that are simultaneously
    object and attribute concept, respectively
M_{spl}: number of supplemental concepts
M_{lbl}: number of labelled concepts, $M_{lbl} := M_{obj} + M_{att} - M_{oa}$

For the sake of avoiding unnecessary concepts, an optimally reduced lattice should have $M_{cpt} \approx M_{lbl}$ if information about supplemental concepts is not available and $M_{cpt} \approx M_{cpt} - M_{spl}$ otherwise. In general, supplemental concepts should be minimised because they do not add any useful information about objects that is not already contained in other concepts. But concepts that are neither object nor attribute concepts and not supplemental should not be removed during reduction because their extensions are important sets according to background knowledge. In order to avoid losing such concepts during reduction, a strategy is to add objects so that $M_{cpt} - M_{spl} = M_{lbl}$. Unfortunately, whether a concept is supplemental can only rarely be deduced from its neighbouring concepts: if a concept a is OR-reducible, then all concepts between a and an attribute concept below it, will be supplemental. In general, if a concept is supplemental, then its lower neighbours should be checked because they could also be supplemental. Sometimes background knowledge provides simple rules about supplemental concepts. For example, if it is known that every object has at least 3 attributes, then any concept at the top of the lattice with fewer than 3 attributes must be supplemental. As demonstrated in the examples below for conceptual partitioning, it can be important to check whether the immediate upper neighbours of object concepts and the joins of object concepts are supplemental because they might be removed during conceptual partitioning.

4   Two Examples of Constructing Euler Diagrams

In this section, the Euler diagrams of two well known examples of lattices are drawn after applying conceptual partitioning. The examples are the body of waters lattice that is on the cover of Ganter and Wille (1999) and the lattice of binary relations (in the same book, after page 85). Ganter and Wille present these lattices as additive or nested line diagrams which highlights Boolean sublattices and symmetries. But based on Chapman et al. (2014), it can be expected that one or two dimensional Euler diagrams will be easier to read for most people.

Figure 3 shows a Hasse diagram for the body of waters example on the left. All of the concepts in the upper half are supplemental (drawn as empty circles) because objects are meant to have 4 attributes. Only object "channel" has 3 attributes because it has not been decided whether it is natural or artificial. In this case, supplemental concepts can be determined by asking users the single question as to whether "channel" is an exception and otherwise objects are expected to have 4 attributes. A goal of a reduction should then be to remove the supplemental concepts. In this example, NOT-reduction might appear promising. But because the set of attributes is a set of 4 antonymic pairs, it is not clear which pair to choose. If NOT-reduction is applied to all 4 pairs, then only the attributes "temporary", "running", "artificial" and "maritime" would be kept. The lattice would consist of very few concepts and most of the supplemental concepts would

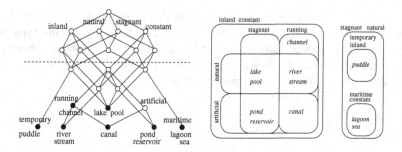

Fig. 3. Body of waters Hasse diagram (left), Euler diagrams "inland AND constant" (middle) and "NOT inland OR NOT constant" (right)

disappear but the lattice would be almost an antichain and not contain much information about the relationships amongst the objects anymore. Therefore after users have seen the results of NOT-reduction, they should be asked whether they would also like to see the result of other reduction methods. In this case a lower horizontal split is possible (the dashed line in Fig. 3) and results in conceptual partitioning using the negatable extension "{inland, constant}". The right half of Fig. 3 shows the resulting partitions "inland AND constant" and "NOT inland OR NOT constant". Implications can still be read from the partitioned Euler diagrams but they need to be quantified. For example: for all objects that are "NOT inland OR NOT constant", inland implies stagnant.

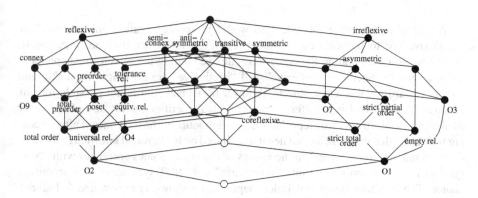

Fig. 4. A concept lattice of types of binary relations

The second example shows a lattice of properties of binary relations (Ganter and Wille 1999). The details of the content of the lattice are not relevant for this paper. A Hasse diagram of the original concept lattice is shown in Fig. 4. In this example, there are only very few supplemental concepts for which Wikipedia can be consulted as a source of background knowledge: if a type of a relation has a name in Wikipedia, then it should not be supplemental and its name can be added as an object. For example, a reflexive and transitive binary relation is called "preorder" in Wikipedia. Although not all supplemental concepts can be decided in that manner, adding objects already has an impact as discussed below. A few of the original objects from Ganter and Wille (1999) do not have special names but must be kept because their object concepts are irreducible. For example, O9 is a connex and antisymmetric relation, but there is no special name for a connex and antisymmetric relation.

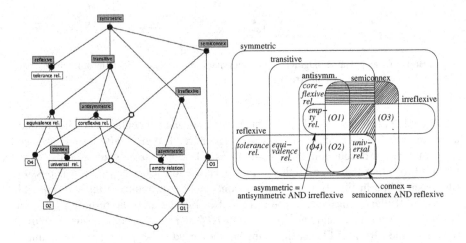

Fig. 5. Hasse and Euler diagram for a concept lattice of symmetric relations

A first obvious choice for reduction would be to use negation on "reflexive" or "irreflexive" or to partition the objects according to whether they are reflexive or irreflexive. But after adding the object "coreflexive" during the search for supplemental concepts, the attributes "reflexive" and "irreflexive" are no longer negatable and no longer partition the set of objects. A lower horizontal split is not possible for this example. A next step is to offer users conceptual partitioning for any of the other attributes whose attribute concepts are lower neighbours of the top concept after checking with users that these are indeed negatable. Based on background knowledge it can be detected that the word "order" in the strings of the object names correlates with "NOT symmetric". Thus partitioning into "symmetric" and "NOT symmetric" is a promising choice. The resulting Hasse and Euler diagrams are shown in Figs. 5 and 6. Figure 5 employs two kinds of shading: horizontal lines for supplemental concepts and diagonal lines for zones that are not concepts of the lattice. The bottom concept has been omitted. If the condition of being a lattice is dropped, then the top 3 shaded zones can be removed leaving a diagram with only 1 shaded zone. Although the Euler diagrams are not showing all of the symmetries of the lattice, they are easy to read. For example, in order to determine the properties of a total order in the Euler diagram of Fig. 6 one only needs to look up (reflexive and semiconnex) and right (transitive and antisymmetric). Optionally below the diagram, it is furthermore stated that connex = reflexive AND semiconnex. Thus connex applies as well.

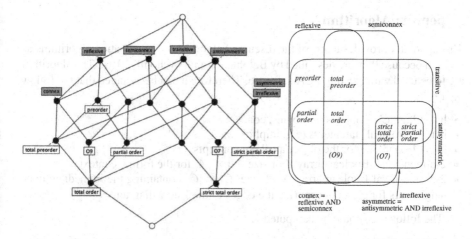

Fig. 6. Hasse and Euler diagram for a concept lattice of relations that are not symmetric

5 Conclusion

This paper discusses means for generating RD-Euler diagrams for representing concept lattices by reducing the information contained in the diagrams, potentially separating the information into several diagrams and implications or other logical expressions. While some algorithms have already been implemented (and are shown in the Appendix), an important future step will be to develop an algorithm for the precise layout of nicely formatted Euler diagrams. A semi-automated algorithm which produces some initial layouts to be further manipulated by users would already be a significant achievement because currently such software appears to be lacking. But a discussion of such algorithms will be left to another paper.

Conducting a user study comparing Hasse and Euler diagrams would also be of interest but would be difficult. Such a study is only sensible if the users have received some training in reading Hasse diagrams. If the study was conducted within the FCA community, then the users might be biased towards Hasse diagrams and furthermore they would have some mathematical background and thus possibly higher spatial reasoning abilities than the average population. FCA knowledge has an impact on reading Hasse diagrams as well. Anybody who observes the Boolean lattice over "total order" in Fig. 6 will instantly read its attributes from the diagram. Anybody who has not been trained in FCA will have to follow all edges that are leading up from "total order". This will be more error prone than reading an Euler diagram if one considers Chapman et al.'s (2014) results. Thus future user studies evaluating numbers of errors in reading RD-Euler diagrams and comparing them to other Euler diagrams might be more insightful than comparing Hasse and Euler diagrams. Since software for producing Hasse diagrams exists, it would be easy to simply add such software to Euler diagram software and give users a choice whether to display Hasse or Euler diagrams.

Appendix: Algorithms

This appendix provides a few of the discussed algorithms. Factorisation algorithms are omitted because they are described by Belohlavek and Vychodil (2010). The algorithms are to be used with existing FCA software. Therefore it can be assumed that the following exist:

- lists O, A, C of objects, attributes, concepts
- list L of logical statements (e.g. implications)
- lists O_C and A_C for the object/attribute concepts
- 2-dimensional Boolean array J of size $|O| \times |A|$ for the formal context
- 2-dimensional Boolean array N of size $|C| \times |C|$ containing the immediate upper neighbours for each concept (i.e. the edges of the Hasse diagram)

The following should be computed:

- 2-dimensional array P of size $|A| \times |A|$ computed by pairwise comparison of the column bitvectors of the attributes. Contains: "e" if the two attributes are equal, "s" for "smaller not equal" (i.e. for every 1 of the first there must be a 1 in the second), "g" for "greater not equal", "n" for negation and "0" otherwise.
- procedure *delete(a)* removes an attribute a from A and J and recomputes everything else as required
- procedure *neigh(c)* returns the set of upper neighbours of concept c according to N
- procedure *meet(a)* returns a (shortest) list of attributes with binary intersection a or NONE.
- procedure *join(a)* returns a (shortest) list of attributes with binary union a or NONE.

The algorithm SynAndOrNot computes attributes that are reducible with Synonym-, AND-, OR- or NOT-reduction. The algorithm Negate computes NEGATION-reduction. Horizontal splits are computed with the final algorithm. For a lower horizontal split, the algorithm HorizontalSplit is used but is stopped as soon as the immediate upper neighbours of all object concepts have been processed. Then the top concepts are removed from each partition and it is checked whether the partitions remain unchanged if all lower neighbours (apart from the bottom concept) are added to each partition.

An Algorithm for Conceptual Partitioning is not provided in detail. Several different steps and strategies can be used to determine which concept extensions are most suitable for partitioning the objects:

- determine which concepts are supplemental and whether objects should be added to non-supplemental concepts
- check whether any other reduction methods are applicable
- run a LOWER HORIZONTAL SPLIT algorithm and determine whether each set of objects corresponds to a concept extension
- any negatable attribute always partitions the set of objects, therefore identify negatable attributes that partition the set of objects approximately in half (can involve single attributes, but also 2 or 3 attributes)
- in order to compare all different possibilities of partitioning, determine which one reduces the number of supplemental concepts
- show different versions to users and ask them which they prefer

Algorithm 1. Algorithm SynAndOrNot

```
 1: for i = 0; i < length(A)-1 do
 2:     if P(A[i],A[i+1]) == "e" then
 3:         ask user to choose one of the two
 4:         a1 = user's choice to be deleted
 5:         delete(a1)
 6:         append "A[i] = A[i+1]" to L
 7:     else if N(A_C(A[i])) contains more than one 1 then
 8:         (A1, ..., An) = meet(A[i])
 9:         delete(A[i])
10:         append "A[i] = A1 AND ... AND An" to L
11:     else if P(A[i],A[i+1]) == "g" and A[i] not in O_C then
12:         (A1, ..., An) = join(A[i])
13:         if not NULL then
14:             ask user whether concept is supplemental
15:             if yes then
16:                 delete(A[i])
17:                 append "A[i] = A1 OR ... OR An" to L
18:     else if P(A[i],A[i+1]) == "n" then
19:         if J(A[i]) contains more than length(O)/2 1s then
20:             a1 = A[i]
21:         else
22:             a1 = A[i+1]
23:         delete(a1)
24:         append "A[i] = NOT A[i+1]" to L
25:     else
26:         next
```

Algorithm 2. Algorithm Negate

```
 1: for i = 0; i < length(A) do
 2:     if J(A[i]) contains more than length(O)/2 1s then
 3:         ask user whether A[i] is negatable
 4:         if yes then
 5:             replace A[i] with NOT(A[i])
 6:             change A and J accordingly
 7:             recompute everything as required
```

Algorithm 3. Algorithm HorizontalSplit

```
1:  initialise Set1 as a list of sets
2:  for item in neigh(bottom_concept) do
3:      Set1[item] = neigh(item)
4:  Cpts = union(neigh(bottom_concept))
5:  to_process = Cpts
6:  processed = {bottom_concept, top_concept}
7:  while length(processed) < length(C) do
8:      for c2 in to_process do
9:          for item1 in Set1 do
10:             if c2 not in Set1[item1] then
11:                 next
12:             for item2 in Set1 do
13:                 if item1 == item2 then
14:                     next
15:                 if c2 in Set1[item2] then
16:                     Set1[item1].add(Set1[item2])
17:                     Set1[item2].add(Set1[item1])
18:      for item1 in Set1 do
19:          for c3 in intersect(Set1[item1],to_process) do
20:              Set1[item1].add(neigh(c3))
21:              Cpts = union(Cpts,neigh(c3))
22:      processed = union(processed,to_process)
23:      to_process = Cpts - processed
24: for item in Set1 do
25:     if length(Set1[item]) < length(C)-2 then
26:         print(Set1[item] - top/bottom_concept is a partition)
```

References

Ainsworth, S.: The functions of multiple representations. Comput. Educ. **33**(2–3), 131–152 (1999)

Alam, M., Le, T.N.N., Napoli, A.: Steps towards interactive formal concept analysis with latviz. In: Proceedings of 5th International Workshop "What can FCA do for Artificial Intelligence"? CEUR Workshop Series (2016). https://ceur-ws.org/Vol-1703/

Belohlavek, R., Vychodil, V.: Discovery of optimal factors in binary data via a novel method of matrix decomposition. J. Comput. Syst. Sci. **76**(1), 3–20 (2010)

Blake, A., Stapleton, G., Rodgers, P., Cheek, L., Howse, J.: The impact of shape on the perception of Euler diagrams. In: Dwyer, T., Purchase, H., Delaney, A. (eds.) Diagrams 2014. LNCS (LNAI), vol. 8578, pp. 123–137. Springer, Heidelberg (2014). https://doi.org/10.1007/978-3-662-44043-8_16

Chapman, P., Stapleton, G., Rodgers, P., Micallef, L., Blake, A.: Visualizing sets: an empirical comparison of diagram types. In: Dwyer, T., Purchase, H., Delaney, A. (eds.) Diagrams 2014. LNCS (LNAI), vol. 8578, pp. 146–160. Springer, Heidelberg (2014). https://doi.org/10.1007/978-3-662-44043-8_18

Cole, R., Eklund, P., Stumme, G.: Document retrieval for e-mail search and discovery using formal concept analysis. Appl. Artif. Intell. **17**(3), 257–280 (2003)

Ganter, B., Wille, R.: Formal Concept Analysis. Mathematical Foundations. Springer, Berlin (1999). https://doi.org/10.1007/978-3-642-59830-2

Ganter, B.: "Properties of Finite Lattices" by S. Reeg and W. Weiß, Revisited. In: Cristea, D., Le Ber, F., Sertkaya, B. (eds.) ICFCA 2019. LNCS (LNAI), vol. 11511, pp. 99–109. Springer, Cham (2019). https://doi.org/10.1007/978-3-030-21462-3_8

Ganter, B., Obiedkov, S.: Conceptual Exploration. Springer, Heidelberg (2016). https://doi.org/10.1007/978-3-662-49291-8

Petersen, W.: Linear coding of non-linear hierarchies: revitalization of an ancient classification method. In: Fink, A., Lausen, B., Seidel, W., Ultsch, A. (eds.) Advances in Data Analysis, Data Handling and Business Intelligence. Studies in Classification, Data Analysis, and Knowledge Organization, pp. 307–316. Springer, Berlin (2009). https://doi.org/10.1007/978-3-642-01044-6_28

Priss, U., Old, L.J.: Concept neighbourhoods in lexical databases. In: Kwuida, L., Sertkaya, B. (eds.) ICFCA 2010. LNCS (LNAI), vol. 5986, pp. 283–295. Springer, Heidelberg (2010). https://doi.org/10.1007/978-3-642-11928-6_20

Priss, U.: Set visualisations with Euler and Hasse diagrams. In: Cochez, M., Croitoru, M., Marquis, P., Rudolph, S. (eds.) GKR 2020. LNCS (LNAI), vol. 12640, pp. 72–83. Springer, Cham (2021). https://doi.org/10.1007/978-3-030-72308-8_5

Priss, U.: Modelling conceptual schemata with formal concept analysis. In: Proceedings of FCA4AI2021 (2021)

Rodgers, P.: A survey of Euler diagrams. J. Vis. Lang. Comput. **25**(3), 134–155 (2014)

Vogt, F., Wille, R.: TOSCANA — a graphical tool for analyzing and exploring data. In: Tamassia, R., Tollis, I.G. (eds.) GD 1994. LNCS, vol. 894, pp. 226–233. Springer, Heidelberg (1995). https://doi.org/10.1007/3-540-58950-3_374

Author Index

D. Dürrschnabel and D. López Rodríguez (Eds.): ICFCA 2023, LNAI 13934, p. 199, 2023.
https://doi.org/10.1007/978-3-031-35949-1

Printed in the United States
by Baker & Taylor Publisher Services